PAS 79:2005

Foreword

This Publicly Available Specification was prepared by BSI in association with C.S. Todd & Associates Ltd with the support and encouragement of the Institution of Fire Engineers and the Northern Ireland Fire Safety Panel, which represents building control and licensing authorities in Northern Ireland, the Northern Ireland Fire Brigade and the Department of Finance and Personnel (DFP) in Northern Ireland. Acknowledgement is given to the following organizations that were consulted in the development of this Publicly Available Specification.

- Bourne Leisure Group Limited
- British Broadcasting Corporation (BBC)
- British Fire Protection Systems Association (BFPSA)
- CB Richard Ellis Limited
- Chief Fire Officers' Association (CFOA)
- C & J Clark Ltd
- Consolidated Assessments
- Edinburgh Fire Consultants Ltd.
- Fire Protection Association (FPA)
- Industry Committee for Emergency Lighting (ICEL)
- Institution of Fire Engineers (IFE)
- Institute of Fire Prevention Officers (IFPO)
- Northern Ireland Fire Safety Panel
- Odeon Cinemas Limited
- Royal Mail Group plc

BSI Committees:

- FSH/14 Fire precautions in buildings
- B/-/12 Fire co-ordination
- HS/1 Occupational health and safety management

This Publicly Available Specification has been developed and published by BSI, which retains its ownership and copyright, except for Annexes A and D (see below). BSI reserves the right to withdraw or amend this PAS on receipt of authoritative advice that it is appropriate to do so. This Publicly Available Specification will be reviewed at intervals not exceeding two years, and any amendments arising from the review will be published as an amendment and publicized in *Update Standards*.

The copyright for Annexes A and D of this Publicly Available Specification is owned by C.S. Todd & Associates Ltd. Purchasers of this Publicly Available Specification are authorized to use the pro-formas contained within these Annexes, and to make an unlimited number of copies for their own use, without infringement of copyright, provided that no changes are made to its text or format.

In England, Scotland and Wales, Regulation 3 of the Management of Health and Safety at Work Regulations 1999[1] requires that all employers carry out a suitable and sufficient assessment of the risks to their employees, and to others, such as to enable the employer to comply with Part II of the Fire Precautions (Workplace) Regulations 1997 (as amended)[2]. Equivalent legislation applies in Northern Ireland [3 and 4]. If the employer employs five or more employees (within the employer's entire organization, regardless of the locations where they are employed), the significant findings of this fire risk assessment must be recorded.

The combination of Part II of the Fire Precautions (Workplace) Regulations (as amended), in conjunction with specific clauses of the Management of Health and Safety at Work Regulations, is known as the "Workplace Fire Precautions Legislation".

At the time of publication of this Publicly Available Specification, the Government intend to use the powers of the Regulatory Reform Act 2001[5] to make major changes to fire safety legislation in England and Wales by means of a new Regulatory Reform (Fire Safety) Order. The effect will be to impose a duty of fire safety care on a defined "responsible person" (such as an employer or managing agent), which will relate to all occupants of the relevant building and to people around the building. The responsible person will have a responsibility to carry out a fire risk assessment to determine the appropriate fire precautions. Proposals exist for similar changes in Scotland and Northern Ireland, via primary legislation.

The fire authority may advise on the fire safety legislation that applies to any building, and on means for compliance. If in doubt regarding the requirements of legislation, consultation with the fire authority is strongly recommended. Advice can also be obtained from a suitably qualified and experienced fire risk assessor or fire safety practitioner.

As a code of practice, this Publicly Available Specification takes the form of guidance and recommendations. It should not be quoted as if it were a specification, and particular care should be taken to ensure that claims of compliance are not misleading.

This Publicly Available Specification is written in practice specification format; commentary on relevant principles is followed by short, succinct recommendations. The purpose of this is to avoid ambiguities and facilitate application by the non-fire specialist. It is envisaged that, when a fire risk assessment is audited for compliance with this Publicly Available Specification, the audit will be based on the recommendations only.

Annex C is normative; all other Annexes are informative.

This Publicly Available Specification is not intended to constitute a textbook on fire safety, and it should not be regarded as a substitute for knowledge of fire safety principles and the practical use and application of fire protection measures. In carrying out the fire risk assessment, there is likely to be a need for reference to other codes of practice and guidance documents on specific aspects of fire prevention, fire protection and management of fire safety, a number of which are listed in the Bibliography.

It has been assumed in the drafting of this Publicly Available Specification that the execution of its provisions will be entrusted to appropriately qualified and competent people.

NOTE Professional bodies may advise on appropriately qualified and experienced specialists. The Institution of Fire Engineers maintain a register of fire risk assessors who have demonstrated to the satisfaction of the IFE that they have a combination of education, training, knowledge and relevant experience in the principles of fire safety, and have experience in carrying out fire risk assessments, one or more samples of which have been subject to review by the IFE to confirm that they are suitable and sufficient to meet their objective.

This Publicly Available Specification does not purport to include all the necessary provisions of a contract. Users are responsible for its correct application.

This Publicly Available Specification is not to be regarded as a British Standard. **Compliance with a Publicly Available Specification does not of itself confer immunity from legal obligations. Attention is drawn to the Workplace Fire Precautions Legislation, guidance on which is contained in *"Fire Safety. An Employer's Guide"*, available from The Stationery Office.**

Contents

	Page
Foreword	ii
Introduction	vi
1 Scope	8
2 Normative references	8
3 Terms and definitions	8
4 The concepts of fire risk and fire hazard	14
5 Principles and scope of fire risk assessments	15
6 Ownership of the fire risk assessment	19
7 Competence of fire risk assessors	21
8 Benchmark standards for assessment of fire precautions	23
9 Documentation of fire risk assessments	25
10 The nine steps to fire risk assessment	26
11 Relevance of information about the building, the occupants and the processes	27
12 Identification of fire hazards and means for their elimination or control	29
13 Assessment of the likelihood of fire	30
14 Assessment of fire protection measures	31
15 Assessment of fire safety management	38
16 Assessment of likely consequences of fire	41
17 Assessment of fire risk	44
18 Formulation of an action plan	45
19 Periodic review of fire risk assessments	47
Annex A (informative) Model pro-forma for documentation of the fire risk assessment	49
Annex B (informative) Fire hazard prompt list	64
Annex C (normative) Key factors to consider in assessment of means of escape	66
Annex D (informative) Model pro-forma for documentation of a review of an existing fire risk assessment	68
Bibliography	75
Other documents	76
Figure 1 — Schematic of fire risk assessment process	17
Figure 2 — Schematic example of appropriate education, training and experience of fire risk assessors	22
Figure 3 — Example of timeline comparison between ASET and escape time (reproduced from BS 7974)	43
Table 1 — A simple risk level estimator	44
Table B.1 — Fire hazards, elimination or control measures and relevant codes of practice	65
Table C.1 — Key factors and specific issues to consider in means of escape	67

© BSI, 2005

Introduction

Employers are required by the Management of Health and Safety at Work Regulations to carry out a "suitable and sufficient" assessment of the fire risks to their employees and others for the purpose of ensuring that the employer complies with the Fire Precautions (Workplace) Regulations 1997. This is usually referred to as a "fire risk assessment". For the purposes of this Publicly Available Specification, a fire risk assessment carried out in accordance with the recommended methodology will be referred to as "the fire risk assessment".

Since these Regulations also require that the assessment is "suitable and sufficient" to ensure that the organization complies with health and safety legislation, the organization could choose to carry out, and document, a single combined health, safety and fire risk assessment. In practice, this approach is normally only ever adopted in the case of very small buildings, and most organizations choose to carry out a separate fire risk assessment, independent of their health and safety risk assessment. The reason for this is that, for most buildings, different skills, experience and expertise are required for each of the two forms of risk assessment.

The term "suitable and sufficient" is not defined in the Regulations. Moreover, the Regulations require that the "significant findings" of the risk assessment, and any group of employees "especially at risk", be recorded if the organization employs five or more people (in the entire organization, and not just in the building in question). Again, the terms "significant findings" and "especially at risk" are not defined in the relevant Regulations. It follows, therefore, that the adequacy of any fire risk assessment is a matter for subjective judgement. This can lead, and has led, to inconsistency in interpretation, creating some difficulties for organizations, their advisers and enforcing authorities. These difficulties have been exacerbated, even for fire safety specialists, by a distinct move, in recent years, towards "risk-proportionate" fire precautions, and away from the more traditional "prescriptive" approach in which there was often a more rigid application of codes of practice without full consideration of fire risk.

This shift is beneficial to those who own and manage buildings, since it provides a better match between risk and precautions, more akin to that found in the field of general health and safety. It therefore precludes unnecessary expenditure in circumstances in which the risk does not justify it. Equally, it ensures adequate protection (possibly to an even higher standard than applied under prescriptive codes) when warranted by the fire risk. Ultimately, the final arbiter as to whether fire precautions satisfy legislation can, however, only be the Courts.

There is, therefore, no single correct or incorrect method of carrying out and recording the significant findings of a fire risk assessment. Rather, there are many approaches that can lead to a suitable, and satisfactorily documented, fire risk assessment, which, at first sight at least, bear little similarity. Nevertheless, the pre-requisites for a suitable and sufficient fire risk assessment are implicit in the Workplace Fire Precautions Legislation, and, accordingly, close scrutiny of most adequate fire risk assessments will reveal consideration of many common factors.

This Publicly Available Specification does not purport to contain a methodology and documentation that is necessarily superior to all others. It is likely to satisfy requirements of forthcoming new fire safety legislation and will, if necessary, be amended at the relevant time to ensure that this is the case. The fire risk assessment methodology is intended to facilitate protection of people from fire. Guidance on fire precautions to protect property, and to protect against interruption to business, from fire may be obtained from property insurers, and many suitably qualified and experienced fire safety consultants may advise on these issues as well as on life safety.

The objectives of this Publicly Available Specification are:

— to provide organizations and their advisers with a methodology for meeting their legislative responsibilities to undertake fire risk assessments;

— to assist non-fire specialists with a framework for assessment of fire risk, albeit that an underpinning knowledge of fire safety principles will be required in order to carry out the fire risk assessment described in this Publicly Available Specification;

— to promote better understanding of fire risks and fire safety by organizations and non-fire specialists;

— to enable common relevant terminology to be adopted by those who carry out fire risk assessments;

— to provide an understanding of the principles and scope of fire risk assessments;

— to establish a pragmatic, holistic and risk-appropriate approach towards assessment of fire prevention measures, fire protection measures and management of fire safety, for the purpose of conducting fire risk assessments;

— to establish a common basis for documentation of fire risk assessments;

— to provide a benchmark for a suitable and sufficient fire risk assessment.

Strictly, the purpose and scope of the Fire Precautions (Workplace) Regulations relate to protection of employees from fire. Currently, therefore, there might be a need for reliance on other legislation to ensure the safety from fire of people other than employees.

However, fire risk assessments carried out in accordance with this Publicly Available Specification address the safety of all occupants of buildings, including employees, visitors, guests and the public (see Clause **1**). There are three reasons for this:

a) In many buildings, the number of employees is small compared to the number of members of the public who could be exposed to risk from fire. It would, therefore, be senseless to carry out a fire risk assessment that ignores the safety of people other than employees.

b) Regulation 3 of the Management of Health and Safety at Work Regulations does refer to assessment of the risk to people other than employees.

c) Future changes to legislation will require fire risk assessments that address the fire safety of all people who occupy or enter a building (and that address maintenance of measures incorporated within a building to assist the fire and rescue service).

1 Scope

This Publicly Available Specification gives one recommended methodology and corresponding documentation for undertaking and recording the significant findings of fire risk assessments in buildings and parts of buildings to which the Workplace Fire Precautions Legislation applies. It is not applicable in the case of domestic dwellings. The methodology is intended to provide a structured approach for people with a knowledge of the principles of fire safety; it is not intended as a guide to fire safety for non-specialists.

The recommended methodology is intended to determine the risk-proportionate fire precautions required to protect building occupants including employees, contractors, visitors and members of the public. The fire risk assessment is not necessarily sufficient to address the safety of fire-fighters, or those outside of the building under assessment, in the event of a fire in the building.

The recommended methodology is not intended to address protection of property (the building and its contents) or the environment, or to address protection of a business, process or activity against interruption. Buildings with special hazards, with the potential for high risk to life (e.g. chemical or nuclear hazards), will require consideration of additional factors associated with these hazards and their means of control are beyond the scope of this document.

2 Normative references

The following referenced document is indispensable for the application of this document. The latest edition of the referenced document (including any amendments) applies.

BS EN ISO 13943, *Fire safety – Vocabulary*.

3 Terms and definitions

For the purposes of this PAS, the terms and definitions in BS EN ISO 13943 and the following apply.

3.1
action plan
measures identified in the course of the fire risk assessment that need to be implemented to ensure that the required level of fire safety is achieved or maintained

NOTE The required standard of fire safety will normally be defined within the organization's fire safety policy, but will never be of a lower standard than required by legislation.

3.2
alarm receiving centre (ARC)
continuously manned premises, remote from those in which a fire alarm system is fitted, where the information concerning the state of the fire alarm system is displayed and/or recorded, so that the fire and rescue service can be summoned

3.3
alternative escape routes
escape routes sufficiently separated by either direction and space, or by fire-resisting construction, to ensure that one is still available should the other be affected by fire

3.4
automatic door release mechanism
device that can be used for holding a door in the open position, against the action of a door closer, and automatically releasing under specified conditions

3.5
available safe egress time (ASET)
time available between ignition of a fire and the time at which tenability criteria are exceeded in a specific space in a building

NOTE To ensure the safety of occupants, the **escape time** (see **3.19**) needs to be shorter than the ASET.

3.6
class A fires
fires involving solid materials, usually of an organic nature, in which combustion normally takes place with the formation of glowing embers

NOTE These are normally carbonaceous fires.

3.7
class B fires
fires involving liquids or liquefiable solids

3.8
class C fires
fires involving gases

3.9
class D fires
fires involving metals

3.10
class F fires
fires involving fats and cooking oils

3.11
combustible
capable of burning in the presence of oxygen

3.12
compartmentation
sub-division of a building by fire-resisting walls and/or floors for the purpose of limiting fire spread within the building

3.13
competent person
person with sufficient training and experience, knowledge or other qualities, to enable him or her to carry out a defined task properly

NOTE This is not necessarily the "competent person" to which Regulation 7 of the Management of Health and Safety at Work Regulations refers.

3.14
dead end
area from which escape from fire is possible in one direction only

3.15
dry rising main (dry riser)
vertical pipe installed in a building for fire-fighting purposes, fitted with inlet connections at the fire and rescue service access level, and with landing valves at specified points, which is normally dry but is capable of being charged with water, usually by pumping from fire and rescue service appliances

3.16
emergency escape lighting
that part of the emergency lighting which is provided to ensure that the escape route is illuminated at all material times

3.17
emergency lighting
lighting provided for use when the supply to the normal lighting fails

3.18
escape route
route forming part of the means of escape from any point in a building to a final exit

3.19
escape time
time from ignition until the time at which all the occupants of the building, or a specified part of the building, are able to reach a place of safety

3.20
evacuation lift
lift that may be used for the evacuation of disabled occupants in a fire under the direction of management or fire-fighters

3.21
false alarm
fire signal resulting from a cause(s) other than fire

3.22
final exit
termination of an escape route from a building, giving direct access to a street, passageway, walkway or open space, where people are no longer in danger from fire

3.23
fire audit
systematic and, whenever possible, independent examination to determine whether standards of fire safety conform to those required in order to achieve the organization's fire safety policy and objectives

3.24
fire damper
mobile closure or intumescent device within a duct, which is operated automatically and is designed to prevent the passage of fire and which, together with its frame, is capable of satisfying for a stated period of time the same fire resistance criterion for integrity as the element of the building construction through which the duct passes

3.25
fire/smoke damper
combined fire and smoke damper

NOTE See **3.24 fire damper** and **3.76 smoke damper**.

3.26
fire door
door or shutter provided for the passage of people, air or objects which, together with its frame and furniture as installed in a building, is intended (when closed) to resist the passage of fire and/or gaseous products of combustion, and is capable of meeting specified performance criteria to those ends

3.27
fire drill (evacuation drill)
rehearsal of the evacuation procedure involving participation of the occupants of the building

3.28
fire equipment sign
safety sign that indicates the location or identification of fire equipment or how it should be used

3.29
fire exposure
extent to which people, animals or items are subjected to the conditions created by fire

3.30
fire hazard
source or situation with potential to result in a fire (e.g. an ignition source or an accumulation of waste that could be subject to ignition)

3.31
fire hazard identification
process of recognizing that a fire hazard exists and defining its characteristics

3.32
fire-fighting lift
lift with fire protection measures, including controls that enable it to be used under the direct control of the fire and rescue service in fighting a fire

3.33
fire load
quantity of heat that could be released by the complete combustion of all the combustible materials in a volume, including the facings of all bounding surfaces

**3.34
fire precautions**
physical, procedural and managerial measures taken to reduce the probability that a fire may occur, and to mitigate the effects of any fire that does occur

**3.35
fire prevention measures**
measures to prevent the outbreak of fire

**3.36
fire procedure**
pre-planned actions to be taken in the event of fire

**3.37
fire protection measures**
design features, systems, equipment or structural measures to reduce danger to people and property by detecting, extinguishing or containing fires

**3.38
fire resistance**
ability of an item to fulfil for a stated period of time the required load-bearing capacity and/or integrity and/or thermal insulation, and/or other expected duty specified in a standard fire resistance test

**3.39
fire risk**
combination of likelihood and consequence(s) of fire

NOTE In the context of this PAS, the relevant consequences are those involving injury to people, as opposed to damage to property.

**3.40
fire risk assessment**
overall process of identifying fire hazards and evaluating the risks to health and safety arising from them, taking account of existing risk controls (or, in the case of a new activity, the proposed risk controls)

**3.41
fire risk assessor**
person who carries out, and documents, a fire risk assessment

NOTE It is essential that the fire risk assessor is a **competent person** (see **3.13**), and the fire risk assessor has a duty of care to the organization on which legislation imposes a requirement for the fire risk assessment. However, the ultimate responsibility for the adequacy of the fire risk assessment rests with that organization, rather than with the fire risk assessor (see Clause **6**).

**3.42
fire safety engineer**
person suitably qualified and experienced in fire safety engineering

**3.43
fire safety engineering**
application of scientific and engineering principles to the protection of people, property and the environment from fire

**3.44
fire safety induction training**
formal training, normally given verbally to new employees, as soon as practicable after their employment, with the objective of imparting sufficient information on the relevant fire risks, fire prevention measures, fire protection measures and fire procedures in the building to ensure the safety of the employee from fire

NOTE Fire safety induction training also assists in preventing the employee from inadvertently putting other occupants of the building at risk from fire.

**3.45
fire safety management**
arrangements to monitor and control fire safety standards, and to ensure that the organization's fire safety policy, once satisfied, continues to be implemented

**3.46
fire safety manager**
person nominated to monitor and control management of fire safety

3.47
fire safety manual
record of all design, procedural and management issues and events that relate to the fire safety of a building

3.48
fire safety objective
specified (or specifiable) goal intended to be achieved by a fire protection measure(s)

3.49
fire safety policy
documented strategy that sets the standards of fire safety an organization is committed to maintaining

NOTE For example, the starting point of a fire safety policy will be that the organization complies with all legislative requirements in respect of fire safety.

3.50
fire safety refresher training
training given to employees periodically to ensure that they remain adequately aware of the fire risks, fire prevention measures, fire protection measures and fire procedures in the building

3.51
fire scenario
detailed description of conditions, including environmental, of one or more stages from before ignition to after completion of combustion in an actual fire at a specific location

3.52
fire stopping
sealing or closing an imperfection of fit between elements, components or constructions of a building, or any joint, so as to restrict penetration of smoke and flame through the imperfection or joint

3.53
fire warden
individual charged with specific responsibilities in the event of fire, normally involving a check to ensure that a particular area of the building has been evacuated

3.54
ignition
initiation of combustion

3.55
ignition source
source of energy that initiates combustion

3.56
inner room
room from which escape is possible only by passing through another ("access") room

3.57
integrity
ability of a separating element, when exposed to fire on one side, to prevent the passage of flames and hot gases or the occurrence of flames on the unexposed side, for a stated period of time in a standard fire resistance test

3.58
maintained emergency lighting
lighting system in which all emergency lighting lamps are illuminated at all material times

3.59
mandatory sign
safety sign that indicates a specific course of action is to be taken

3.60
manual call point
component of a fire detection and fire alarm system that is used for the manual initiation of a fire alarm signal

3.61
material alteration
alteration that changes (usually lowering) the standard of fire protection originally provided

3.62
means of escape
structural means whereby (in the event of fire) a safe route or routes is or are provided for people to travel from any point in a building to a place of safety (without external assistance)

3.63
non-combustible
not capable of undergoing combustion under specified conditions

3.64
non-maintained emergency lighting
lighting system in which all emergency lighting lamps are illuminated only when the supply to the normal lighting fails

3.65
occupant(s) at special risk
building occupant(s) who, as a result of their physical or mental state, age or location in the building, are at greater risk from fire than an able-bodied, fully alert adult afforded adequate means of escape and other fire precautions, whether on a short-term or long-term basis

3.66
panic bolt
mechanism consisting of a minimum of two sliding boltheads that engage with keepers in the surrounding door frame or floor for securing a door when closed; the mechanism can be released by hand or body pressure on a bar positioned horizontally across the inside face of the door

3.67
panic latch
mechanism for securing a door when closed; the latch bolt can be released by hand or body pressure on a bar positioned horizontally across the inside face of the door

3.68
phased evacuation
system of evacuation in which different parts of the building are evacuated in a controlled sequence of phases, those parts of the building expected to be at greatest risk being evacuated first

3.69
place of safety
place in which people are in no danger from fire

3.70
products of combustion
solid, liquid and gaseous materials resulting from combustion

3.71
protected (corridor, route or staircase)
corridor, route or staircase enclosed in fire-resisting construction

3.72
refuge
area that is enclosed with fire-resisting construction (other than any part that is an external wall of a building) and served directly by a safe route to a storey exit, evacuation lift or final exit, thus constituting a temporarily safe space for disabled occupants to await assistance for their evacuation

NOTE Refuges are relatively safe waiting areas for short periods. They are not areas where disabled occupants should be left indefinitely until rescued by the fire and rescue service or until the fire is extinguished. It is the organization's responsibility to provide assistance, and the arrangements for this should be incorporated within the building's fire procedures.

3.73
responsible person
person on whom legislation imposes a requirement for the fire risk assessment

NOTE The responsible person is normally an organization, such as an employer, rather than a specific named person.

PAS 79:2005

3.74
safe condition sign
safety sign that provides information about safe conditions (e.g. a fire exit sign)

3.75
smoke alarm
device containing within one housing all the components, except possibly the energy source, necessary for detecting smoke and for giving an audible alarm

NOTE The term "smoke alarm" is normally reserved for devices intended for domestic use.

3.76
smoke damper
mechanical device which, when closed, prevents smoke passing through an aperture within a duct or structure

NOTE The device may be open or closed in its normal position and may be automatically or manually actuated.

3.77
structural fire protection
features in layout and/or construction that are intended to reduce the effects of a fire

3.78
third party fire risk assessor
fire risk assessor who is not an employee of the responsible person (e.g. a consultant)

3.79
tolerable (fire risk)
of a level acceptable to the organization, taking into account the requirements of fire safety legislation, the fire safety policy of the organization (see **3.49**), the nature of the building, the fire hazards in the building (see **3.30**), the nature of the occupants, the cost of additional fire precautions and any other relevant factors

3.80
travel distance
actual distance to be travelled by a person from any point within the floor area to the nearest storey exit, having regard to the layout of walls, partitions and fixings

3.81
voice alarm system
sound distribution system that provides means for automatically broadcasting speech messages and warning signals

3.82
wet rising main (wet riser)
vertical pipe installed in a building for fire-fighting purposes and permanently charged with water from a pressurized supply, fitted with landing valves at specific points

4 The concepts of fire risk and fire hazard

4.1 Commentary

*It is important that, in the fire risk assessment, confusion does not result from loose, inexact or conflicting use of terminology. Particular care needs to be taken to avoid improper use of the terms "fire hazard" (see **3.30**) and "fire risk" (see **3.39**). BS 8800 defines a hazard as a source or a situation with a potential for harm in terms of death, ill health or injury, or a combination of these. Accordingly, in this document, a fire hazard is defined as a source or situation with potential to result in a fire. Thus, the presence of uncontrolled fire hazards affects the likelihood of fire, rather than the consequences of fire.*

*Since a fire risk assessment and a health and safety risk assessment could, in fact, form part of a single risk assessment (see **Introduction**), it is logical that the concepts of risk and fire risk are consistent with other fields of risk assessment. Accordingly for the purpose of this PAS, fire risk is defined as the combination of the probability of fire occurring and the magnitude of the consequences of fire (see **3.39**).*

NOTE This definition differs from that in BS 4422-1 (which defines "fire risk" as the risk of fire occurring) but is consistent with the definition in BS EN ISO 13943. It is also consistent with usage in the general safety field, by organizations such as the Health and Safety Executive, and the concept of risk used in BS 8800.

This clear distinction between fire hazard and fire risk is of great value in any analytical approach to fire safety, but particularly in a fire risk assessment. It can be considered that fire risk is the product of multiplying the probability of fire by a measure of the consequences of fire if it does occur. Thus, for example, even though the likelihood of fire occurring might be low, the fire risk could still be high as a result of potential for serious injury to occupants in the event of fire. For example, the potential for serious injury could result from inadequate provision of fire exits and/or inadequate means of giving warning to people in the event of fire. Such circumstances would be likely to be regarded intuitively, even by a layman, as high risk, and accordingly this definition of fire risk is likely to be relatively intuitive even to non-fire specialists.

4.2 Recommendations

The following recommendations are applicable.

a) In the fire risk assessment, care should be taken to distinguish clearly between the concepts of fire hazard and fire risk.

b) In the fire risk assessment, the terms "fire hazard" and "fire risk" should only be used in a context consistent with the definitions given in sub-clauses **3.30** and **3.39**.

5 Principles and scope of fire risk assessments

5.1 Commentary

*The fire risk assessment is a structured assessment of the fire risk (see Clause **4**) in the relevant building for the purpose of expressing the current level of fire risk, determining the adequacy of existing fire precautions (see **3.34**) and determining the need for, and nature of, any additional fire precautions. Any such additional fire precautions required are set out in the action plan (see **3.1**), which forms part of the documented fire risk assessment (see Clause **9**). The objective of the action plan is to set out measures that will ensure that the fire risk is reduced to, or maintained at, a tolerable level (see **3.79**).*

The fire risk assessment needs to be a genuine and open-minded approach to the assessment of fire risk and fire precautions. It is not, for example, appropriate to use the fire risk assessment to justify a decision regarding fire precautions that has already been made, or to justify significant departures from universally recognized good practice.

*It follows from the definition of fire risk that the fire risk assessment involves consideration of relevant fire hazards and the means for their elimination or control, i.e. fire prevention measures. This contrasts with the approach adopted in most traditional fire safety legislation, which tends to concentrate on **fire protection** measures (see **3.37**), rather than fire prevention measures (see **3.35**).*

This approach to fire risk assessment tends to parallel that adopted in health and safety risk assessments, whereby the objective of the risk assessment is not limited to merely preventing harm to people as a result of a hazard, but begins with endeavours to eliminate or reduce the hazard itself. Thus, the fire risk assessment begins with endeavours to reduce the likelihood of fire. In this sense alone, fire risk assessment is a more holistic approach to control of fire risk than that traditionally adopted under legislation.

*The likelihood of fire can, however, never be reduced to zero. Accordingly, there is normally need for fire protection measures of the nature commonly prescribed under traditional legislation, such as means of escape (see **3.62**), measures that assist in the use of escape routes (see **3.18**), means of giving warning of fire and means for fighting fire. However, fire protection measures, by definition, only have a bearing on fire safety after fire has occurred and, therefore, fire prevention has failed.*

*Most of the visible fire precautions in a building are fire protection measures, and it is with these measures that the fire safety provisions within building regulations are primarily concerned. However, in a modern building, the risk to people (and property) from fire is often governed more by the quality of fire safety management (see **3.45**) than the level of fire protection. Indeed, significant factors in most non-domestic, multiple fatality fires, particularly those involving, say, 10 deaths or more, are failures in fire safety management, rather than failures in building design or fire protection measures.*

Thus, in contrast with the approach to compliance with building regulations, it is absolutely essential that every fire risk assessment gives thorough attention to fire safety management and, therefore, to matters such as fire procedures, staff training, testing and maintenance of fire protection equipment, inspection of means of escape, etc. Good fire safety management also contributes to the prevention of fire by incorporating policies and measures that reduce the likelihood of fire.

It follows, therefore, that the fire risk assessment can only validly be carried out on a building that is in use, so that the actual working conditions, practices and procedures can be taken into account. The fire risk assessment is not a means for snagging fire precautions in a newly constructed building prior to occupation.

NOTE Parts of the fire risk assessment can be used for such a purpose, in order to ensure the building is safe for occupation, but such an exercise would not constitute a suitable and sufficient fire risk assessment, as management issues and operational issues cannot be properly addressed.

The fire prevention measures, fire protection measures and components of fire safety management can be considered as variables, the standard of which can be reduced or increased, according to the fire risk, in order to provide an integrated package of measures that limits fire risk to a tolerable level. However, some factors that have major impact on fire risk are not variable, but are "given" factors for the building in question.

Such factors include:

 a) *the height of the building (e.g. single storey or multi-storey, low rise or high rise, the presence of basements);*

 b) *the construction of the building (e.g. largely non-combustible (see **3.63**) or mainly combustible (see **3.11**));*

 c) *the activities and processes carried out in the building (e.g. handling of highly flammable materials, creation of combustible wastes, use of ignition sources (see **3.55**));*

 d) *the complexity of the building (e.g. simple, straightforward escape routes or complex, convoluted escape routes);*

 e) *the floor area of each floor;*

 f) *the nature of the occupants (e.g. young or old, infirm or able-bodied);*

 g) *the familiarity of the occupants with the building (e.g. fully familiar, slightly familiar or totally unfamiliar);*

 h) *the state (or likely state) of the occupants (e.g. awake or asleep, alert or under the influence of alcohol or drugs);*

 i) *the history of fires in the building.*

Although the above factors cannot (or cannot readily) be changed, their effect on fire risk (primarily as a result of their effect on the consequences of a fire) needs to be taken into account in the fire risk assessment, so that they are reflected in the level of fire risk expressed in the fire risk assessment. The level of fire precautions then needs to be tailored to the level of risk.

*Since the likelihood (i.e. probability) of fire and the consequences of fire, if it does occur, are largely independent factors in the fire risk assessment (see Clause **4**), they need to be considered separately in the fire risk assessment (see Figure 1). For example, in a single-storey, open plan building with an abundance of readily available fire exits, a high probability of fire (e.g. as a result of numerous small fires in an industrial process) does not imply serious consequences to occupants (in terms of injury) in the event of fire. On the other hand, in a large, multi-storey building with minimal fire load (see **3.33**) and few ignition sources (e.g. a store for metal components), if there is inadequate means of escape and inadequate means of warning people in the event of fire, the consequences to occupants in the event of fire could be serious. It should, equally, be noted that, in each of these examples, poor standards of fire safety management could affect both the probability of fire and the consequences of fire.*

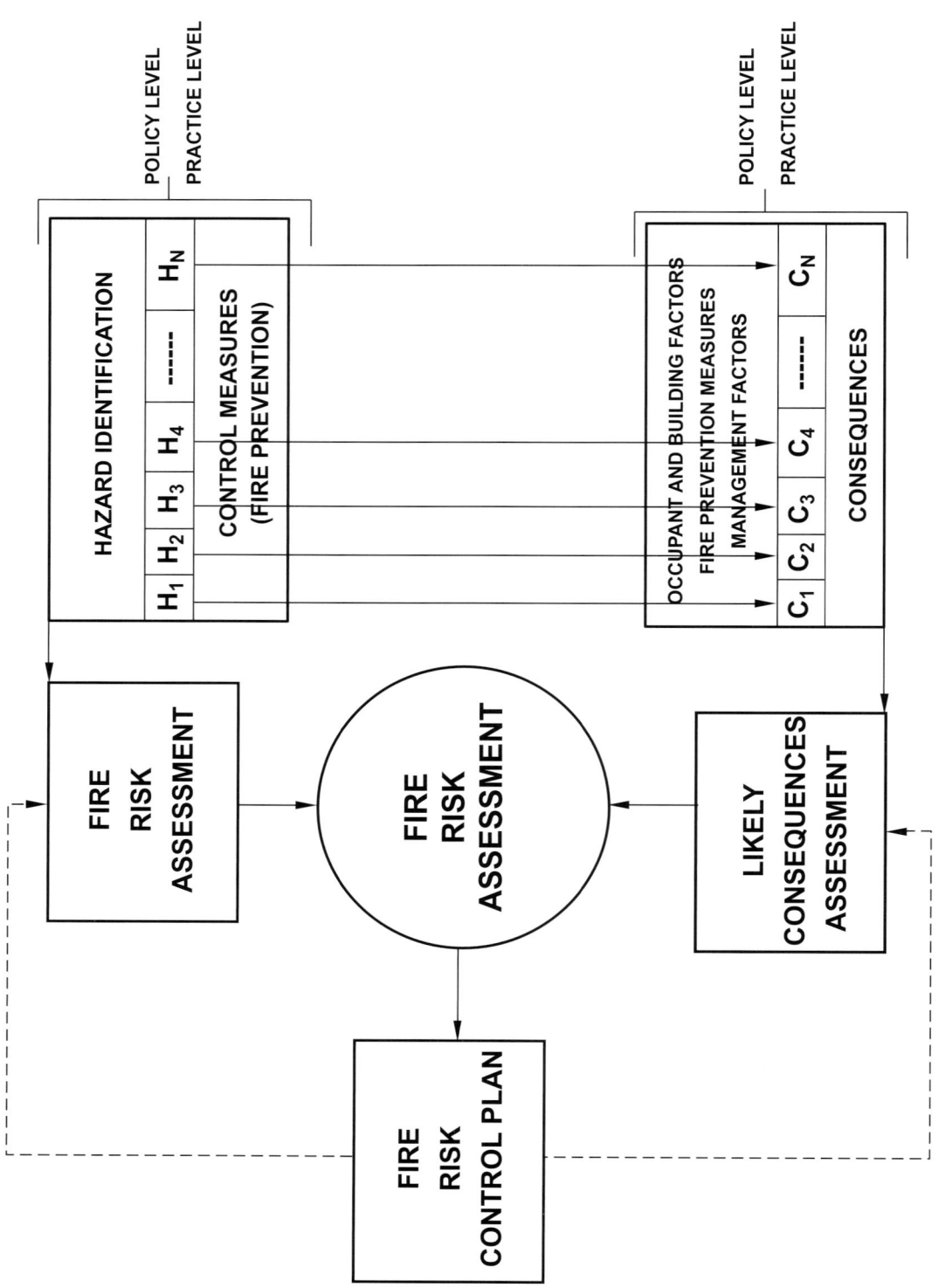

Figure 1 — Schematic of fire risk assessment process

Once the level of fire risk is determined, any need for improvements in fire precautions can be identified. The separate consideration of probability of fire and consequences of fire is then of value, since, if the fire risk is unacceptably high, the source(s) of the high fire risk can be identified by separating the fire risk into its two component factors. It can then be determined whether the problem is primarily one of high likelihood of fire, necessitating fire prevention measures in the action plan, or serious consequences in the event of fire, necessitating fire protection measures, or a combination of the two.

*The determination of the likelihood of fire, the consequences of fire, and hence the fire risk, can normally be subjective in nature, and will not normally be quantified numerically. Numeric methods, including calculation of probabilities and use of fire scenarios (see **3.51**), need normally only be used in specialist industries (such as the chemical industry) with potential for very high fire risk, or be used in the formulation of designs based on complex fire safety engineering (see **3.43**). Moreover, care is necessary to ensure that simple points schemes, which purport to evaluate fire risk numerically, are not misleading (see Clause **17**).*

Where the original design of a building has been based on fire safety engineering and approved under relevant building regulations, it is not generally necessary to check this design from first principles in the course of the fire risk assessment. It is, however, necessary to ensure that features and facilities that form part of the design are being properly maintained and managed.

*The action plan (see Clause **18**) needs to contain measures that are practicable and risk-proportionate, while resulting in compliance with legislation and the organization's fire safety policy (see **3.49**). The nature of the measures specified needs to be such that they are likely to receive acceptance by management and other occupants who may be affected by them.*

5.2 Recommendations

The following recommendations are applicable.

a) The fire risk assessment should only be carried out when a building is in normal use. If, in the case of a new or refurbished building, there is a need to carry out a fire risk assessment before the building is fully occupied and in normal use, a further fire risk assessment should be carried out once the building is in normal use.

b) Every documented fire risk assessment should explicitly set out information on the following matters:

  1) the height of the building, or part of the building, that is the subject of the fire risk assessment, or the number of storeys above and below ground;

  2) brief details of construction, with information about any aspects that make a significant contribution to risk;

  3) the activities and processes carried out in the building;

  4) approximate number of occupants;

  5) whether the building will be occupied by members of the public (as opposed to employees), and, if so, the approximate number of members of the public (if known);

  6) approximate floor area of the building, or a typical floor of the building, or part of the building, that is the subject of the fire risk assessment;

  7) in the case of a building in multiple occupation, the nature of other occupancies (if known);

  8) occupants at special risk (see **3.65**) in the event of fire (e.g. sleeping occupants, disabled occupants, those working in remote areas, etc);

  9) any fires that have occurred in recent years (if known);

  10) any further relevant information that has a bearing on fire risk (other than the information described in recommendation **5.2 c)**) or on the validity of the fire risk assessment.

c) Within every documented fire risk assessment, it should be clear that proper consideration has been given to the following matters, regarding which there should be, at least, basic information and, where relevant, comment:

   1) fire hazards and means for their elimination or their control;
   2) fire protection measures;
   3) relevant aspects of fire safety management.

d) Every documented fire risk assessment should contain an expression of the level of fire risk, determined from the information specified in recommendations **5.2 b**) and **5.2 c**).

   NOTE The level of fire risk may normally be expressed subjectively (e.g. trivial, tolerable, moderate, substantial, intolerable).

e) Every documented fire risk assessment should contain an action plan, unless it is expressly confirmed within the fire risk assessment that no additional fire precautions are necessary.

6 Ownership of the fire risk assessment

6.1 Commentary

Regardless of whether the fire risk assessment is carried out by, for example, staff of an organization, or by a third party fire risk assessor (see **3.78**)*, the ultimate responsibility for the adequacy of the fire risk assessment rests with the "responsible person" defined by legislation as responsible for ensuring that the fire risk assessment is carried out and that the fire precautions are adequate.*

NOTE 1 For example, under the Workplace Fire Precautions Legislation, this person is the employer.

In some buildings, such as commercial buildings in multiple occupation, the landlord or managing agent can also have responsibilities under legislation, and, in order to discharge their duty, they might also need to carry out a fire risk assessment to determine the adequacy of those fire precautions for which they are responsible (e.g. fire precautions in the common parts and fire protection systems that are building-wide, such as the building's fire alarm system).

It is important that the responsible person understands and acknowledges their responsibility for the fire risk assessment, as this, again, is in contrast with traditional legislation), with which the responsible person might be more familiar.

NOTE 2 Examples of legislation espousing the traditional approach include the Fire Precautions Act and the Fire Services (Northern Ireland) Order.

Under traditional legislation, the principal responsibility for judging the adequacy of existing fire precautions, and for determining the requisite additional fire precautions, rested with the fire authority. This has led to an inappropriate attitude of reliance on the fire authority by those who own, operate or occupy buildings. This attitude is out of keeping with the modern principle in the health and safety field that those who create risks (e.g. by virtue of operating a building or processes in the building) have prime responsibility for taking action to mitigate them. Thus, application of prescriptive norms for fire precautions by the fire authority (e.g. as a pre-requisite for issue of a fire certificate) has proved not to be conducive to pro-active fire safety management, which is the key to effective control of fire risk. There has, for example, been a tendency for some building occupiers to perceive a fire certificate as an end in itself, rather than a vehicle for securing continuing safety of occupants from fire.

The fire risk assessment constitutes an underpinning for an organization's self-compliance with fire safety legislation and the organization's fire safety policy. It is essential that the organization does not treat the fire risk assessment as a mere formality, or treat the documented fire risk assessment as a formal document that is an end in itself and that is simply stored away until the fire authority request sight of it.

Such an attitude would be akin to that adopted by less responsible organizations towards a fire certificate, but might arise, in particular, if the fire risk assessment is carried out on behalf of the organization by a third party fire assessor, such as a consultant. Since a third party can only make recommendations in the action plan, but cannot enforce requirements, the fire risk assessment process would then suffer from the disadvantages of the fire certification process, without the accompanying advantage of the latter process, namely the positive enforcement of measures identified as necessary. Self-compliance would then simply provide a framework for prosecuting organizations when shortcomings in fire precautions were identified by the fire authority, or, even worse, after serious injury or loss of life occurred as a result of a fire.

*Where, within an organization, there is a competent person (see **3.13** and Clause **7**), able to carry out the fire risk assessment, it is appropriate for that person to carry out, or oversee any third party that carries out, the organization's fire risk assessments. If fire risk assessments are carried out by a third party, such as a consultant, it is essential that the organization for whom the fire risk assessment is carried out understands the role of the third party; the role is to facilitate the fire risk assessment and to advise on fire precautions, but the responsibility for the adequacy of the fire risk assessment and adequacy of fire precautions rests with the organization. It is the responsibility of the organization to ensure that whoever carries out the fire risk assessment is competent (see Clause **7**), as its ownership remains with the organization.*

Where the fire risk assessment is carried out for an organization by a third party, it is essential that the organization buys into the fire risk assessment from the outset. This means that the organization will need to provide information and support for whoever carries out the fire risk assessment, as much of the essential information required in order to carry out the fire risk assessment will reside within the organization and cannot be obtained by a third party without the organization's co-operation.
*The organization will also need to give practical support to the fire risk assessor (see **3.41**) by ensuring that the fire risk assessor has access to appropriate people from whom information must be obtained and has sight of relevant documentation, and by facilitating access to all areas of the building.*

Where practicable, to ensure acceptance of the action plan, the recommendations in the action plan need, in the course of the fire risk assessment, to be discussed with the management of the building in question to ensure that the documented fire risk assessment is delivered to the appropriate person(s), namely the person(s) on whom the findings impact and who can arrange for implementation of the action plan.
*The "ownership" of the fire risk assessment will then continue throughout the life of the building, so that, for example, the fire risk assessment is subject to review at an appropriate frequency and when changes take place (see Clause **19**).*

6.2 Recommendations

The following recommendations are applicable.

a) Where legislation imposes a requirement on any organization for a fire risk assessment to be carried out, it should be clearly understood by the organization that the responsibility for the adequacy and accuracy of the fire risk assessment, and of the information contained therein, rests with that organization, rather than the fire risk assessor, regardless of whether the fire risk assessor is an employee of the organization or a third party (e.g. a consultant).

b) Where, within an organization, an employee of the organization is competent to carry out the fire risk assessment, where practicable that person should carry out, or oversee any third party that carries out, the organization's fire risk assessment.

c) The organization should take all reasonable steps to ensure that every fire risk assessor who carries out fire risk assessments on behalf of the organization is competent to carry out this task, regardless of whether the fire risk assessor is an employee of the organization or a third party, such as a consultant (see Clause **7**).

   NOTE Certain professional bodies maintain a register of people who have demonstrated, to the satisfaction of the professional body, competence in carrying out fire risk assessments. Registration is available to those who only carry out fire risk assessments within their own organization, as well as to those who offer this as a commercial service. Use of a person who is listed, by name, on such a register satisfies recommendation **6.2 c**) and might constitute evidence that the organization took reasonable steps to ensure the competence of the fire risk assessor.

d) The organization should ensure that the fire risk assessor has access to appropriate people and relevant documentation, is provided with all relevant information and has access to all areas of the relevant building, or part of the building, at the time of the fire risk assessment, particularly if the fire risk assessment is carried out by a third party.

e) The documented fire risk assessment should be studied carefully by appropriate people in the organization to confirm the accuracy of documented information, understand the contents, particularly the fire hazards and any shortcomings in fire protection measures or fire safety management, and to implement the action plan.

f) After the fire risk assessment has been carried out, it should be subject to periodic review, particularly when changes that could affect fire risk occur or when there is any other reason to suspect that the fire risk assessment is no longer valid (see Clause **19**).

7 Competence of fire risk assessors

7.1 Commentary

*For many buildings, the fire risk assessment, and its periodic review (see Clause **19**), will be the sole underpinning for continued adequacy of fire precautions on an ongoing basis, after compliance with building regulations. It is, therefore, essential that fire risk assessments are only carried out by a competent person (see **3.13**).*

Competence does not necessarily depend on the possession of specific qualifications, although such qualifications might contribute to the demonstration of competence. In the case of small simple buildings, where the fire risk assessor might, for example, be an employee of the occupier, it is possible that the following attributes of the fire risk assessor might be sufficient in conjunction with a study of suitable guidance documents:

   a) an understanding of relevant current best fire safety practices in buildings of the type in question;

   b) an awareness of the limitations of the fire risk assessor's own experience and knowledge;

   c) a willingness and ability to supplement existing experience and knowledge, when necessary, by obtaining external help and advice.

Larger buildings will require a higher level of knowledge and experience on the part of the fire risk assessor. For complex buildings, there will be a need for the specific applied knowledge and skills of an appropriately qualified specialist. In such cases, evidence of specialist training and experience, or membership of a professional body, can enable competence to be demonstrated.

In general, other than in the case of simple, low risk buildings, fire risk assessors, particularly those offering their services on a commercial basis (e.g. consultants), need:

   1) a good understanding of the legislation under which the fire risk assessment is required;

   2) a sound underpinning combination of education, training, knowledge and experience in the principles of fire safety;

   3) an understanding of fire development and the manner in which people behave when exposed to fire;

   4) training and/or experience in carrying out fire risk assessments;

   5) an understanding of the fire hazards, fire risks and occupants at special risk from fire that are likely to occur in the building, or part of the building, for which the fire risk assessment is carried out.

In the context of the above paragraph, education is likely to involve formal education of a relatively academic nature, often culminating in a qualification (although not necessarily to degree level). Training involves training of a practical nature, often given on the job. Knowledge can be obtained by academic study, training, working alongside others, short courses, continuing professional development or any combination of two or more of these.

It is not implied that education, training and experience in the principles of fire safety need each be extensive, provided that the combination of each results in adequate knowledge. Moreover, a high level in respect of any one of these might compensate for a lower level in another. This is shown diagrammatically in Figure 2.

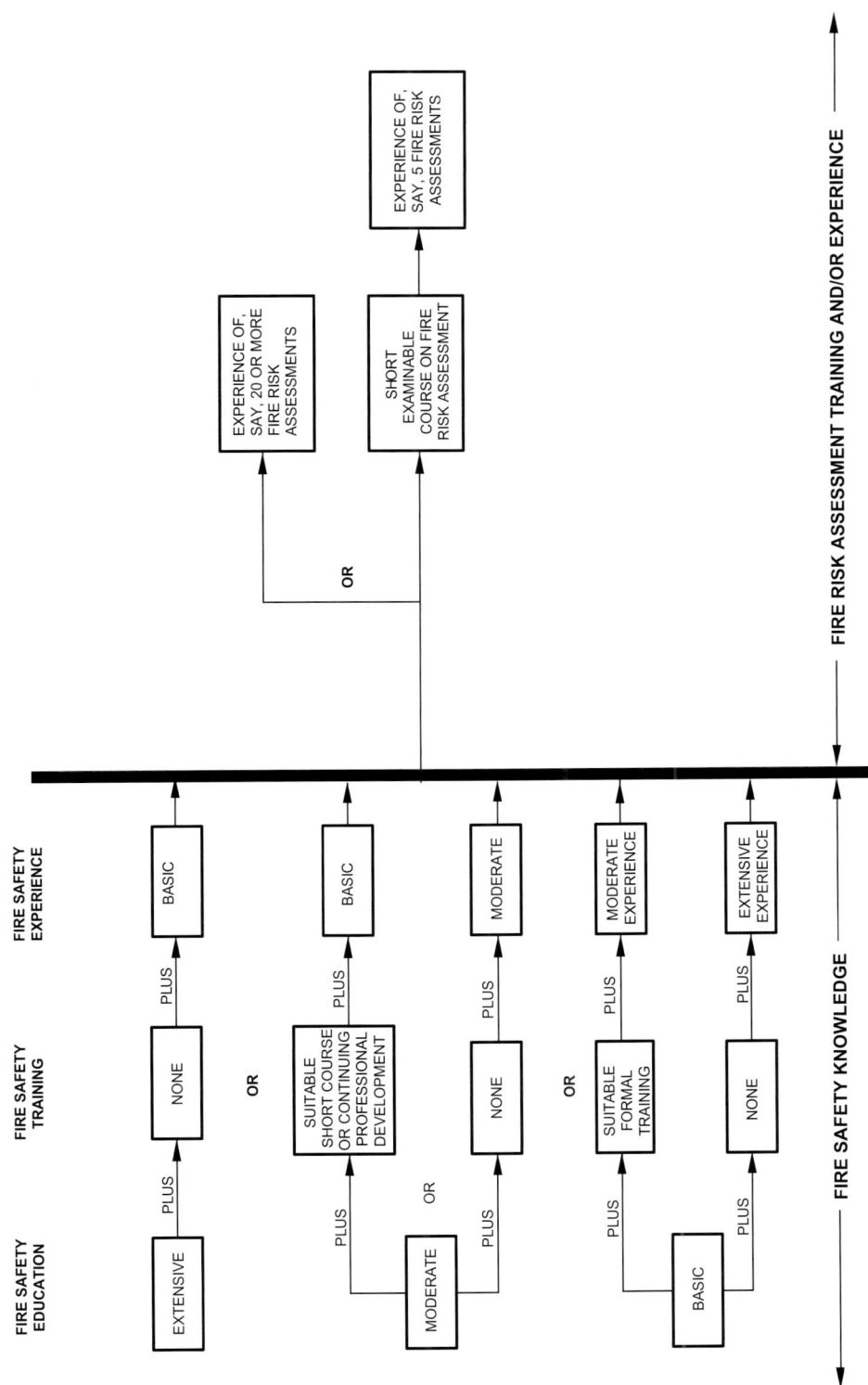

Figure 2 — Schematic example of appropriate education, training and experience of fire risk assessors

In Figure 2, fire safety education that is extensive might, for example, comprise education to HNC level or above, while moderate education might comprise a formal course carried out over several weeks, or in several modules. Basic fire safety education might comprise a course of just a few days duration, or education given as part of academic education in a relevant discipline (e.g. a degree in building surveying). Basic experience might involve the practice of fire safety for a period of at least six months, while moderate experience might be experience of, say, three years duration. Extensive experience might be no fewer than, say, eight years.

Often, fire risk assessors' education, training and, particularly, experience has related to relatively rigid application of prescriptive codes of practice, with only minimal opportunity to exercise professional judgement that would result in risk-proportionate fire precautions, above or below the standards of those prescribed in the relevant code of practice. A minor amount of training and/or experience might then be necessary to convert their competence in the principles of fire safety to competence in fire risk assessment (see Figure 2). For example, successful completion of a short, examinable course, plus experience of carrying out, say, five fire risk assessments over a period of, say, three months might be sufficient. Alternatively, experience of carrying out a greater number of fire risk assessments (say, 20 or more) over a longer period of time (say, six months) might be equally sufficient.

7.2 Recommendations

The following recommendations are applicable.

a) All fire risk assessments should be carried out by a competent person (see **3.13**).

b) The fire risk assessor need not possess any specific academic qualifications but should:

1) understand the relevant fire safety legislation;

2) have appropriate education, training, knowledge and experience in the principles of fire safety;

3) have an understanding of fire development and the behaviour of people in fire;

4) understand the fire hazards, fire risks and relevant factors associated with occupants at special risk within buildings of the type in question;

5) have appropriate training and/or experience in carrying out fire risk assessments.

8 Benchmark standards for assessment of fire precautions

8.1 Commentary

The assessment of fire precautions in the fire risk assessment does not normally merely involve rigid comparison of existing fire precautions with standards set out in prescriptive codes of practice. Similarly, the action plan is not based on rigid adherence to prescriptive norms found in codes of practice. To adopt such an approach would not necessarily result in risk-proportionate fire precautions.

Nevertheless, in assessing or formulating measures to eliminate or control fire hazards, it will often be appropriate, in the case of certain fire hazards, such as potential electrical faults, to adopt guidance in recognized codes of practice. This will particularly be the case where these codes of practice are well established, universally recognized, produced by authoritative bodies with specialist knowledge regarding the hazard in question, and are based on sound scientific or engineering principles (as opposed to arbitrary judgements).

Thus, for example, in considering the fire hazard created by defective electrical wiring, it will normally be appropriate to control the hazard by inspection and testing of the fixed electrical installation in accordance with BS 7671 and with guidance produced on this subject by the Institution of Electrical Engineers. It would normally be inappropriate for the fire risk assessor to advocate control measures that conflict with such guidance.

However, in the case of other fire hazards, such as smoking or control of combustible waste, the knowledge, experience and judgement of the fire risk assessor will be much more important. Although there is ample guidance on such matters in various publications, the guidance is less universally recognized, more general in nature and not exactly applicable in every situation.

In the case of fire protection measures, a plethora of codes of practice exist. In the case of some specific fire protection systems, a single, universally accepted code of practice exists and is based on sound engineering principles. This is the case in respect of, for example, fire detection and fire alarm installations, emergency escape lighting installations and automatic sprinkler installations. These codes of practice are invariably adopted in the design of new installations.

In such codes of practice, certain parameters specified in the codes, such as the sound pressure levels of a fire alarm system or the illuminance levels of emergency escape lighting, are acknowledged to be relatively arbitrary in nature. Thus, minor variations from numerically expressed limitations or performance levels need not necessarily have any significant effect on fire risk. Nevertheless, where the action plan includes recommendations for upgrading any aspect of the relevant system (e.g. improvement in the sound pressure levels of a fire alarm system or the illuminance levels of an emergency escape lighting installation), it will be appropriate to adopt the relevant recommendations of the appropriate code of practice within the action plan.

Traditionally, the design of various other fire protection measures, specified within the relevant code of practice, is often based more on custom and practice, and on arbitrary recommendations, than on scientific and engineering principles. Moreover, often various conflicting recommendations occur within different codes of practice on the same subject. In addition, sometimes different recommendations apply to new and existing buildings. For example, recommendations within guidance that supports building regulations often differ from recommendations within guidance that supports legislation applicable to existing buildings, such as the Fire Precautions Act. This makes rigid adherence to any particular code of practice even less appropriate.

*A classic example of this concerns means of escape (see **3.62**). For example, different maximum travel distances (see **3.80**) are recommended in different codes of practice dealing with different buildings, and even in different codes of practice that can be applied to the same building. Yet, travel distance is a fundamental component in the design of means of escape. Similar variations exist in the more detailed recommendations of various codes of practice.*

This has led to a school of thought amongst some experts that the application of prescriptive codes of practice within the fire risk assessment is inappropriate. However, while there is a need for risk-proportionate fire precautions, rather than rigid application of prescriptive norms, it should be borne in mind that prescriptive codes of practice have achieved their objective; for example, it is rare for multiple fatality deaths to occur in non-domestic buildings that comply with the relevant prescriptive code of practice, unless a number of failures in fire safety management have occurred.

This might be as a result of the continual development of codes of practice over many years, and of the fact that, when a code of practice has been found to be deficient following a major fire disaster, the code of practice has been "patched" to address the deficiency. After many revisions and "patches", prescriptive codes of practice arguably result in a level of fire precautions that is sufficient to reduce fire risk to a tolerable level, and in fire protection measures that are relatively forgiving in the event of inadequate fire prevention measures and shortcomings in fire safety management.

However, although it has always been intended that codes of practice be flexible in their application, there is a perception (sometimes, but not always, correctly) that there has been inflexible application of codes of practice. This has arguably tended to result in unnecessary restrictions on the design and use of buildings, and in over-extensive fire precautions.

The "one size fits all" nature of prescriptive codes can also result in lower standards of fire protection measures than warranted by the fire risk. An example of this is the assumption within many codes of practice that automatic fire detection is never necessary in buildings in which no one sleeps, other than as compensation for reduction or variation in the standards of other fire protection measures, to operate other fire protection measures or for protection of inner rooms. However, the fire risk assessment might well determine that there is a need for some automatic fire detection in such a building. It should also be noted that prescriptive standards can become outdated and fire protection measures designed in accordance with such standards might not be sufficient.

*At the design stage of a building, the alternative to application of all recommendations within a prescriptive code of practice is the application of fire safety engineering (see **3.43**), usually in conjunction with many, but not all, of the recommendations from the codes of practice. However, formulation of fire protection measures from a first principles approach to fire safety, for example using fire safety engineering, is complex, time consuming and demands the expertise of specialists, such as a fire safety engineer (see **3.42**). It is not usually an appropriate approach to the fire risk assessments required by legislation, albeit that the principles of fire safety engineering, applied subjectively, can be relevant.*

*For example, when fire occurs, a key factor in the safety of occupants is the escape time (see **3.19**). Control of maximum travel distance and minimum exit widths, using the same figures for all buildings of the same purpose group, is an imprecise way of ensuring that escape time is suitably limited, and only addresses the time between response of occupants to an alarm signal and the point at which they reach a place of relative safety (often described as "evacuation time"). This approach ignores time for detection of fire, the subsequent time interval before an alarm signal is given to occupants and the time for* *occupants to recognize the alarm signal. Moreover, it takes no account of the time for occupants to respond to the fire alarm signal (which might be longer than the combination of all other time intervals and the evacuation time).*

*However, calculation or prediction of these time intervals is extremely difficult. Furthermore, a knowledge of escape time in isolation is of little value. It is more appropriate to compare escape time with the ASET (see **3.5**), which is the time between ignition and the occurrence of untenable conditions that would result in serious injury or death of occupants.*

This document is intended to be suitable for use by, for example, fire risk assessors with a background in application or enforcement of traditional prescriptive fire protection codes of practice. Accordingly, it is assumed that these codes of practice will be a starting point or benchmark for assessment of the adequacy of fire precautions in the building. It is, however, further assumed that the fire risk assessor is capable of exercising judgement to determine whether the recommendations of prescriptive codes of practice should be relaxed, or added to, in order to determine the appropriate level of fire precautions and to formulate a risk-proportionate action plan. Appropriate guidance is given in subsequent clauses.

8.2 Recommendations

The following recommendations are applicable.

a) Assessment of fire precautions should take into account guidance within relevant, recognized codes of practice, albeit that rigid adherence to these might not be necessary. While fire precautions recommended in the action plan should also take account of such codes of practice, the recommendations in the action plan should be risk-proportionate, which may necessitate measures of a standard below or above that specified in the relevant code of practice.

b) Departures from the recommendations of recognized codes of practice should be based on the judgement of the fire risk assessor, and should take into account relevant fire safety, or fire safety engineering, principles.

NOTE It is of benefit, particularly to those who subsequently audit the fire risk assessment, such as enforcing authorities, if significant departures from recognized codes of practice, deemed acceptable by the fire risk assessor, are recorded and, preferably, justified in the documented fire risk assessment (see Clause **9**).

9 Documentation of fire risk assessments

9.1 Commentary

There is no single correct means of documenting the fire risk assessment, nor are there specific, definitive requirements within legislation for the content of a documented fire risk assessment, only that the "significant findings" and any group of occupants especially at risk be recorded. The fire risk assessor therefore needs to make a judgement as to what constitutes "significant findings" and occupants especially at risk.

*See **5.2** for the information that needs to be taken into account in the fire risk assessment and the matters on which judgements need to be made. Such information needs to be documented, along with other relevant factual information (e.g. managerial responsibility for fire safety). In the case of certain matters, particularly the "given" factors taken into account in assessment of the fire risk (see **5.1**), information about the factors (e.g. number of storeys of the building) needs to be recorded.*

*In the case of other matters, such as certain fire protection measures (e.g. emergency escape lighting), it might be sufficient to acknowledge that proper consideration has been given to the matter (see **5.2 c**)), without necessarily recording descriptive information about it. Indeed, unnecessary detail might not be conducive to ensuring that the relevant responsible person(s) study the document properly or take note of significant findings. However, for example, if the standard of a fire precaution (such as means of escape) departs significantly from a recognized norm, but the departure is considered acceptable by the fire risk assessor, it is of value to document the justification for this.*

*Annex A contains a pro-forma that is considered "suitable and sufficient" means for documenting the fire risk assessment. The pro-forma contained in Annex A is only a model, in that, if completed by a competent person (see Clause **7**), the scope of the documented fire risk assessment will comply with the recommendations of this PAS. Equally, the format of a documented fire risk assessment may vary from that shown in Annex A, provided that all recommendations are satisfied.*

9.2 Recommendations

The following recommendations are applicable.

NOTE 1 Use of the pro-forma in Annex A enables compliance with the recommendations of this sub-clause, but, equally, other formats may be used, provided that, as a minimum, the information and matters included in the pro-forma in Annex A are addressed.

a) A documented fire risk assessment should comply with recommendations **5.2 b) – 5.2 e)** of this PAS.

b) If any fire protection measure obviously and significantly departs from the standard recommended in a relevant prescriptive code of practice, but no upgrading of the measure is recommended in the action plan, the acceptance of the existing standard should be justified within the documented fire risk assessment.

NOTE 2 The departures to which **9.2 b)** refers are primarily those affecting provisions for means of escape and functional aspects of fire protection systems; it is not, for example, intended that justification of the continued use of a fire alarm system or emergency escape lighting system designed in accordance with a superseded standard would normally be necessary.

c) The fire risk assessment should record the name of the fire risk assessor(s), the date(s) on which the fire risk assessment was carried out and the name(s) of the principal person(s) with whom there was consultation (e.g. for supply of relevant information) at the time of the fire risk assessment.

d) The fire risk assessment should record any significant areas of the building to which access was not possible at the time of the fire risk assessment.

e) The fire risk assessment should record the date by which it should be subject to review (see Clause **19**).

10 The nine steps to fire risk assessment

10.1 Commentary

To promote a structured approach to fire risk assessment (see Clause 5), there are nine steps in the ongoing fire risk assessment process, somewhat akin to the five steps to risk assessment often adopted in a health and safety risk assessment, and to the eight steps described in BS 8800.

*a) The first step is to obtain relevant information about the building, the processes carried out in the building, and about the occupants of the building. Information about previous fires is also of value, particularly where the organization has multiple sites with common operations. This information was described in Clause **5** as the "given" factors in the fire risk assessment. Much of the relevant information will usually be obtained by interviewing a relevant representative(s) of the management, prior to carrying out a physical inspection of the building.*

*b) The second step is fire hazard identification (see **3.31**) and the determination of measures for the elimination or control of the identified fire hazards. This will normally involve a combination of interviewing the management and inspection of the building.*

c) The third step is to make a (subjective) assessment of the likelihood of fire. This will be based primarily on the findings of step two (see Figure 1). However, the assessment of the likelihood of fire will also take into account any relevant information obtained in step one.

*d) The fourth step is to determine the physical fire protection measures (see **3.37**), relevant to protection of people in the event of fire. The relevant information can, again, be obtained partly from the initial discussion with management, but will, primarily, be obtained by inspection of the building, so that the standard of fire protection can be determined.*

*e) The fifth step is to determine relevant information about fire safety management (see **3.45**). This will, primarily, involve discussion with management, but might also involve examination of documentation, such as records of testing, maintenance, training, drills, etc.*

*f) The sixth step is to make a (subjective) assessment of the likely consequences to occupants in the event of fire (see Figure 1). This assessment needs to take account of the fire risk assessor's opinion of the likelihood of various fire scenarios (see **3.51**), the extent of injury that could occur to occupants in these scenarios, and the number of people affected. This assessment is principally based on the fire risk assessor's findings in steps four and five, but will take account of information obtained in the first step.*

*g) The seventh step is to make an assessment of the fire risk and to decide if the fire risk is tolerable (see Figure 1). The fire risk is assessed by combining the likelihood of fire and the consequences of fire (see Clause **17**).*

*h) The eighth step is to formulate an action plan (see **3.1**), if this is necessary to address shortcomings in fire precautions in order to reduce the fire risk. Even if fire risk is assessed as tolerable, there might be a need for minor improvements in fire precautions. See Clause **18** for formulation of an action plan.*

*i) Thereafter, in the ninth step, the fire risk assessment is subject to periodic review (see Clause **19**). Review of the fire risk assessment is necessary after a period of time defined in the fire risk assessment, or at an earlier time if changes take place, or if there are other reasons to suspect that the fire risk assessment is no longer valid.*

The nine steps set out above, while in a logical, structured order, are not necessarily set out in the chronological order in which the steps are carried out on site. For example, some information relevant to control of fire hazards, the determination of fire protection measures and the management of fire safety is normally most appropriately obtained in a single meeting that is held prior to inspection of the building.

10.2 Recommendations

The following recommendations are applicable.

a) In all fire risk assessments carried out in accordance with this PAS, the fire risk assessor should explicitly take the following nine steps:

NOTE Explicitly, in the above context, means that in the documented fire risk assessment it should be clear that each of the nine steps has been taken by the fire risk assessor.

1) obtain information on the building, the processes carried out in the building and the people present, or likely to be present, in the building;

2) identify the fire hazards and means for their elimination or control;

3) assess the likelihood of fire, at least in subjective terms;

4) determine the fire protection measures in the building;

5) obtain relevant information about fire safety management;

6) make an assessment of the likely consequences to people in the event of fire, at least in subjective terms;

7) make an assessment of the fire risk;

8) formulate and document an action plan, with prioritization if appropriate;

9) define the date by which the fire risk assessment should be reviewed.

b) The fire risk assessment should be reviewed after a period of time defined in the fire risk assessment, or such earlier time as significant changes take place or there are other reasons to suspect that the fire risk assessment is no longer valid (see Clause **19**).

11 Relevance of information about the building, the occupants and the processes

11.1 Commentary

*Clause **5.1** sets out various "given" factors that have a major impact on fire risk. It is relevant to document information about these factors in the fire risk assessment. The manner in which the factors should be taken into account in the fire risk assessment process is discussed in this current sub-clause.*

Firstly, the number of floors below ground and the number of floors above ground need to be determined. In assessing the fire risk, it needs to be borne in mind that basements can present particular difficulties for fire-fighting and, hence, rescue. Mitigating factors would be, for example, low population within basement floors, occupation only, or primarily, by trained staff, and the presence of fire protection measures, such as automatic sprinkler protection, automatic fire detection and means for removal of smoke.

Deep basements can result in somewhat prolonged evacuation times for occupants, as will also be the case in high buildings. In the latter case, rescue by the fire and rescue service is more difficult from floors above the height of normal fire and rescue service ladders and even more difficult in the case of very tall buildings with floors beyond the reach of a turntable ladder or hydraulic platform. The time for occupants to descend staircases can be quite significant. Again, robust protection of staircases, automatic fire detection and automatic sprinkler protection are mitigating factors.

Floor area, on each floor, is also a relevant factor to consider. Evacuation of a very large floor area is likely to take longer than evacuation of a much smaller floor area, and the number of occupants is likely to be greater. Similarly, complex escape routes might take longer to negotiate than simple ones. Consideration also needs to be given to the construction of the building. This can have an effect on fire development, particularly if combustible building construction is likely to be involved in the fire prior to evacuation of occupants.

The general use to which the building is put (the occupancy) is also relevant. From a knowledge of occupancy, conclusions can normally be drawn regarding the activities carried out, the nature and state of occupants, whether members of the public are in the building, etc. These are relevant factors in the assessment of fire risk.

A further important consideration is the maximum number of occupants that can reasonably be expected at any one time. It is important that the number recorded in the fire risk assessment is a reasonably foreseeable maximum, so that it forms a basis for any calculations of required exit capacity, etc (see Clause 14).

Where practicable, it is of value for the number of occupants to be sub-divided into employees and members of the public. Apart from the relevance of this information to the fire risk assessment, for certain occupancies, the number of employees present at any one time will influence the need, or otherwise, for a fire certificate. With regard to the fire risk assessment, employees can be, and are more likely to be, trained in procedures to follow in the event of fire, and they are normally familiar with the building. On the other hand, members of the public are likely to be unfamiliar with the building and will not have received any formal instruction in fire procedures, etc. This has a bearing on the likely consequences in the event of fire. The ratio of staff to members of the public can also have a bearing on the effectiveness of evacuation procedures. Other occupants to whom consideration might be necessary include cleaners, contractors, visitors, etc.

Particular account needs to be taken of occupants who could be at special risk in the event of fire.

NOTE Attention is drawn to the Management of Health and Safety at Work Regulations' requirements regarding the recording of occupants at special risk in the event of fire.

Particular care needs to be taken to ensure proper consideration of disabled occupants, who are often at special risk in the event of fire in view of the possible need for assistance with evacuation or special warning of fire. All forms of disability need to be considered, including mobility impairment, deafness, blindness and learning difficulties.

It is also appropriate to regard sleeping occupants as at special risk in the event of fire. They are less likely to be aware of the fire, might not be roused immediately by the fire alarm signal, might be disorientated when first aroused from sleep (particularly if under the influence of alcohol or drugs) and might be reluctant to evacuate. Generally, in buildings that incorporate sleeping accommodation, there will be a need for a high standard of automatic fire detection and emergency escape lighting (see Clause 14).

It is also possible that occupants working alone in remote areas of the building could be at special risk in the event of fire. Their location at the time of a fire might be unknown to other building occupants, and there might be no one to assist them if they were trapped by the fire or overcome by smoke. If, for example, people were working on the roof of the building, the fire alarm signal might not be sufficiently audible and their means of escape might be restricted.

Other occupants at special risk include any occupants for whom immediate escape might not be possible, who might not be adequately warned of fire, etc. For example, it is not unknown for cleaners, or others, working in a building during the night, to have restricted means of escape, which might not be acceptable.

The most important purpose of considering and recording occupants at special risk in the event of fire is to ensure that adequate provisions are in place to protect such occupants from fire. Having recorded such occupants within the fire risk assessment, it needs to be clear within a documented fire risk assessment that there are provisions to ensure the safety of these occupants.

It is relevant to take account of any fire, however small, that is known to have occurred within recent years. Factors to consider include the circumstances of the fire, including the cause, and any remedial action take to prevent a reoccurrence. Information of this type can be of use in identifying fire hazards that would not, otherwise, be obvious from an inspection of the building. Where the fire risk assessment is carried out throughout a large number of buildings under the control of one organization, review of the fire loss experience at each building can sometimes reveal significant trends or identify remedial action that might be appropriate throughout all buildings to rectify a latent hazard.

11.2 Recommendations

The following recommendations are applicable.

   a) In carrying out the fire risk assessment, the fire risk assessor should take account of the information described in **5.2 b)**.

   b) While it is not normally necessary to document, within the fire risk assessment, the manner in which every factor to which **11.2 a)** refers affects fire risk or has been taken into account, there should normally be explicit information within the fire risk assessment regarding appropriate measures to protect occupants who are at special risk in the event of fire.

12 Identification of fire hazards and means for their elimination or control

12.1 Commentary

*In this step of the fire risk assessment, the fire risk assessor identifies all reasonably foreseeable and significant fire hazards and examines the measures in place for their elimination or control. By definition, this means considering potential ignition sources (see **3.55**) and situations that have the potential to result in a fire. It is necessary, therefore, for the fire risk assessor to be aware of the common causes of fire in the type of building under assessment, and to have an understanding of the work processes in the building under assessment, as well as an understanding of the policies and procedures that contribute to prevention of fire. At the conclusion of this step of the fire risk assessment, the fire risk assessor will be in a position to assess the likelihood of fire (see Clause 13).*

*It is assumed that the fire risk assessor is already familiar with the common causes of fire and is either aware of recognized good practice in the elimination or control of fire hazards (i.e. is aware of recognized fire prevention measures (see **3.35**)), or has access to appropriate codes of practice. Normally, the documented fire risk assessment comprises a pro-forma, which incorporates a prompt list of fire hazards that need to be considered in the fire risk assessment. A suitable prompt list of fire hazards, typical measures for their elimination or control, and relevant codes of practice that give further guidance, is set out in Annex B.*

The list of fire hazards in the prompt list in Annex B is not necessarily exhaustive, and other fire hazards might need to be considered, particularly those relating to specific work processes carried out in the building. For example, significant ignition sources, perhaps associated with mechanical, chemical or electrical processes, might be identified in the fire risk assessment, and care needs to be taken to ensure that any unacceptable practices or measures for control of such ignition sources are identified, and, where relevant, are recorded within the documented fire risk assessment (see Clause 9). It might also be appropriate to record relevant control measures. On the other hand, fire hazards with negligible potential for harm need not be documented or given further consideration.

*It should also be noted from the definition of fire hazard (see **3.30**) that fire hazards are not limited to ignition sources per se. Various situations can constitute fire hazards. For example, combustible storage or rubbish does not, in itself, constitute a source of ignition. However, if, for example, the storage or rubbish is positioned close to the windows of a building, it might be ignited maliciously, or accidentally by discarded smokers' materials, and the resultant fire could then spread into the building via the windows. Such a situation would, therefore, constitute a fire hazard.*

It is often appropriate to consider the means for control or elimination of fire hazards in two distinct phases, which can be regarded as policy and practice. For example, in the case of the fire hazard created by faulty electrical appliances, one control measure might be a policy that portable electrical appliances are subject to periodic inspection and testing. The "practice" stage comes when the building is inspected and observations can be made as to whether there is adherence to the policy. It might then be found that, for example, some appliances are overlooked in the programme of inspection and testing, or it might be noted that some staff bring their own electrical appliances, such as radios, heaters, fans, etc, into the workplace, without these appliances being subject to inspection or test.

12.1 Recommendations

The following recommendations are applicable.

   a) The fire risk assessment should give consideration to means for elimination or control of, at least, the common causes of fire, and shortcomings in such measures should be addressed within the action plan (see Clause **18**).

b) Specific causes of fire that should be considered in every fire risk assessment include arson, electrical faults, smoking, cooking (if any is carried out), inadequate control over the use of portable heaters, contractors' activities and "hot work", inadequate maintenance of heating installations, and lightning.

   NOTE It is possible that there will be a need for consideration of other fire hazards, including those associated with work processes and energy-using appliances.

c) Consideration of fire hazards should not be limited to those comprising specific sources of ignition. Situations, such as poor housekeeping, that could lead to a fire should also receive proper consideration.

13 Assessment of the likelihood of fire

13.1 Commentary

Once all relevant fire hazards have been identified, and measures for their control or elimination have been determined, the fire risk assessor is in a position to make an assessment of the likelihood of fire. It would be possible, in theory, to associate a likelihood of fire with each of the identified fire hazards. However, this would make the fire risk assessment process unnecessarily complex and unduly lengthy. Usually, it is sufficient to consider the overall likelihood of fire in the building; this may be regarded as the summation of likelihoods of fire associated with each and every one of the fire hazards identified.

The likelihood of fire need not, and usually cannot, be expressed in a meaningful numeric manner, such as in terms of a statistical probability of fire. All that is required is a subjective judgement that classifies likelihood of fire into one of several pre-determined categories. Since the assessment of these factors is subjective, the use of numbers to express likelihood of fire does not confer any greater accuracy to the assessment of fire risk.

The pre-determined categories of likelihood of fire may be described in the form of words, such as "low", "medium" and "high" or in the form of numbers (e.g. 1, 2 and 3), but there will be a need for at least three categories. However, if likelihood is expressed in the form of numbers, care is necessary to ensure that it is not implied, for instance, that a likelihood of "2" indicates that fire is twice as likely to occur compared to a likelihood of "1".

There is no upper limit to the number of categories of likelihood that can be adopted in the fire risk assessment process. However, if too many categories are adopted, the distinctions between categories will be meaningless. Moreover, if the same fire risk assessment process is then applied to numerous different buildings (e.g. within the estate of a single organization), particularly by different fire risk assessors, assessments of likelihood of fire are likely to be inconsistent, and the benefits of comparing the fire risk in different buildings (e.g. for the purpose of prioritizing improvements on a building-by-building basis) will then be lost.

*If likelihood of fire is judged to be typical for buildings of the type in question, it is normally appropriate to ascribe to the building the middle category of the pre-determined categories of likelihood of fire. Higher categories can then be used to indicate serious shortcomings in elimination or control of fire hazards (i.e. fire prevention), while lower categories can be used in cases where the likelihood of fire is abnormally low (e.g. because the building is secure and not normally occupied). Minor shortcomings in fire prevention measures need not be regarded as changing the category ascribed to the building, but need to be addressed in the action plan (see Clause **18**).*

13.2 Recommendations

The following recommendations are applicable.

a) In the process of every fire risk assessment, an assessment should be made of the likelihood of fire. It is usual and acceptable for the likelihood of fire to be expressed subjectively (e.g. "low", "normal" or "high").

b) If, in the fire risk assessment methodology adopted, likelihood of fire is expressed in terms of one of several pre-determined categories, the number of pre-determined categories should be an odd number, so that the middle category can be adopted for buildings that are typical for buildings of the type and occupancy in question. The number of pre-determined categories should be at least three, but may be more than three.

14 Assessment of fire protection measures

14.1 Commentary

14.1.1 *General*

In this step of the fire risk assessment, consideration is given to those physical measures incorporated within the building that are intended to mitigate the consequences of fire (and, hence, limit fire risk) in terms of harm to occupants of the building in the event of fire. These measures are, by definition, fire protection measures, and their effect is to limit fire exposure (see 3.29).

Once fire occurs, the first requirement is to warn people, who can then use suitably designed means of escape (see 3.62). To use the means of escape safely and efficiently, there will often be a need for suitable signs and for emergency escape lighting (see 3.16). Harm to occupants might also be mitigated, and safe escape facilitated, by appropriate measures to control or extinguish the fire (whether by use of portable fire extinguishers or hose reels by occupants, or by activation of an automatic fire suppression system, such as an automatic sprinkler system).

It follows, therefore, that the fire risk assessor needs to take account of, and assess the adequacy of, the following fire protection measures:

a) the fire warning system in the building and the means for its activation (e.g. manual operation or a combination of manual and automatic means for activation);

b) the means of escape from the building;

c) any relevant, or required, means for facilitating safe and efficient use of means of escape (such as signs and emergency escape lighting);

d) structural and similar measures for limiting fire development;

e) means for fighting fire;

f) other relevant fire protection systems.

Adequacy of the engineering design, installation and commissioning of fire protection systems can often be certified by organizations that are themselves third-party certificated as competent in their specialist field. More generally, there is a need for all fire protection systems to be commissioned, designed, installed and maintained by engineers competent in this specialist field.

Fire development and spread can be passively limited by fire protection measures (see 3.37), such as fire resisting walls and floors (over and above any required to protect means of escape), which can be used to sub-divide the building into a number of separate fire compartments (e.g. to satisfy the requirements of building regulations for compartmentation (see 3.12)). It will often be relevant, therefore, for the fire risk assessor to take account of such fire resisting construction and to consider its maintenance (e.g. the adequacy of fire stopping (see 3.52)), often by inspecting sample areas of construction. The spread of fire can also be actively limited by means for fighting fire, such as fire extinguishing appliances or automatic fire suppression systems.

On the other hand, fire development can be assisted by, for example, flammable linings on walls or ceilings, or by readily flammable furniture, furnishings, and by the accumulation of combustible material, including waste material. The fire risk assessor needs to take into account the presence and location of these features, and, sometimes, their physical state. For example, damage to upholstered furniture can result in exposure of foam fillings, which might be easily ignited and result in rapid development of fire.

Since the earliest effect of fire on building occupants is often loss of visibility on escape routes as a result of smoke, there will also be a need to take account of measures to limit spread or build up of smoke. These can range from fire doors (see 3.26) to active smoke control systems, such as those designed to extract smoke or to maintain a positive pressure within escape routes to prevent the ingress of smoke.

In the sub-clauses that follow, the key fire protection measures that affect the consequences of fire are considered separately. The factors are not, however, independent. In assessing the likely consequences of fire (see Clause 16), a judgement needs to be made regarding the effect of each of the fire protection measures discussed below, and of a number of the management issues discussed in Clause 15, on the escape time (see 3.19) or on the ASET (see 3.5).

14.1.2 Fire detection and warning

The arrangements for detection of fire and the means for then warning occupants of the building need to be considered. Fire can be detected by people or by automatic fire detectors. If people are present in the area of fire origin, they normally detect fire before it is detected automatically by, for example, smoke or heat detectors.

Traditionally, therefore, automatic fire detection is only considered necessary in the following buildings and situations:

    a) buildings in which people sleep (e.g. hotels, boarding houses, hostels, residential care buildings and hospitals);

    b) covered shopping complexes, and in large or complex places of public assembly;

    *c) in buildings with phased evacuation (see **3.68**);*

    *d) as compensation for reduction in the standards of certain other fire protection measures below the norms prescribed in prescriptive codes of practice (e.g. extended travel distances (see **3.80**) or reduction in the fire resistance of construction protecting escape routes);*

    *e) in lieu of vision between an inner room (see **3.56**) and its associated access room;*

    f) as a means of operating other fire protection systems (e.g. automatic closure of fire doors, automatic release of electronically locked doors, or initiation of smoke control systems).

In general, therefore, automatic fire detection has not, traditionally, been considered necessary in common places of work where no one sleeps (e.g. offices, shops, factories and warehouses) and most other non-residential buildings (e.g. libraries, schools and community buildings), except for the purpose of property protection. Normally, if automatic fire detection is not required for compliance with current building regulations, it is unusual (but not unknown) for the fire risk assessment to identify a requirement for automatic fire detection. For example, this may arise from a low level of occupancy of an area of the building, from which there could be extensive fire spread before detection.

NOTE 1 Attention is drawn to guidance that supports national building regulations (e.g. in England and Wales, Approved Document B [7] or the relevant part of BS 5588) and to guidance that relates to existing buildings (e.g. *Fire Safety. An Employer's Guide* [6]).

*Notwithstanding the above, since the fire risk assessment does not involve rigid application of prescriptive codes of practice (see Clause 8), it is appropriate to consider whether particular circumstances dictate the need for automatic fire detection in buildings in which such detection was, traditionally, deemed unnecessary. Such circumstances might be related to likely shortcomings in the reliability of management standards or fire procedures, levels of surveillance that are unusually low for an occupancy of the type in question, or processes that constitute an abnormal fire hazard for the occupancy. For example, even if people detect a fire before an automatic fire detector does so, there is often a delay before they operate the fire alarm system, albeit that the delay might be reduced by training and fire drills (see Clause **15**); automatic fire detection might then reduce the overall delay between ignition and giving warning to occupants.*

Moreover, since the fire risk assessment involves a holistic assessment of fire precautions, rather than independent prescription of a number of fire protection measures in isolation of consideration of other fire protection measures, there is sometimes scope, in the fire risk assessment, for greater use of automatic fire detection to compensate for existing standards of structural fire precautions that are lower than the prescribed norm.

Where automatic fire detection is considered necessary, the areas in which it is installed, and the types of detection used, need to take into account the objective of the fire detection and the importance of avoiding false alarms. A system that produces too many false alarms can result in a reduction in the level of fire safety, as people then become reluctant to evacuate when the evacuation signal is given.

For example, in buildings in which people sleep, the primary objective is normally to ensure that occupants, other than a person in the room of fire origin, are given sufficient warning to use the escape routes before smoke within the escape routes makes escape impossible. Normally, this necessitates the provision of smoke detectors within the escape routes themselves, and the provision of virtually any form of fire detector (e.g. heat, smoke or carbon monoxide detectors) in rooms that open onto escape routes. Smoke detectors within these rooms would provide earlier detection that heat detectors (and would offer protection to the occupant of the room of fire origin), but heat detectors often suffice for protection of those beyond the room of fire origin. In some occupancies (e.g. student halls of residence), the provision of smoke detectors, rather than heat detectors, in bedrooms might cause an untoward level of false alarms, such that, in providing enhanced protection for the occupant of the room of fire origin, complacency as a result of false alarms would create a risk to other occupants. It is appropriate for the fire risk assessment to take account of this.

Systems installed prior to 2002 might not comply in full with current recommendations, particularly in respect of certain aspects of engineering design. In many cases, this is perfectly acceptable, but it is appropriate for new systems and new work associated with the modification of existing systems, recommended in the action plan, to comply with current recommendations.

NOTE 2 Guidance on types of fire detectors, their application and limitation of false alarms is given in BS 5839-1.

NOTE 3 Normally, domestic smoke alarms (see **3.75**) are unsuitable for non-domestic buildings, and any use of these devices would need to be fully justified in the fire risk assessment.

*If the fire risk assessment considers the provision of automatic fire detection, to compensate for standards of other fire protection measures that are below the relevant norm, the fire risk assessor needs to ensure that early detection is sufficient to compensate for this. As a minimum requirement, this will require a subjective consideration of likely fire scenarios (see **3.51**). In such a case, the fire risk assessor will normally need significant experience in the practice of fire safety, or might need specialist advice. The automatic fire detection will need to comply with current recommendations, at least in terms of types of fire detectors and their location. The fire risk assessor will need to specify carefully the areas of the building in which any additional detectors should be provided.*

NOTE 4 The relevant recommendations are those in BS 5839-1 referring to Category L5 systems, which are those where the areas in which automatic fire detection is provided are tailor made to achieve a specific fire objective.

*Most buildings in which automatic fire detection is not required need a manual ("break glass") electrical fire alarm system, so that the fire alarm can be raised by anyone who discovers a fire. Only in very small buildings will word of mouth (i.e. shouting "Fire") or mechanical devices, such as rotary gongs, be a sufficient means of giving warning to other occupants. As part of the fire risk assessment, it needs to be determined whether the number and sitting of manual call points are sufficient, on the basis that it should not be possible to leave any storey of the building, or leave the building by means of a final exit (see **3.22**) without passing at least one manual call point.*

*Consideration might need to be given to the nature of fire warning signals. Usually, these are given by bells or electronic sounders. However, a voice alarm system (see **3.81**) might be more appropriate, or even necessary, in some buildings, such as those in which the public are present in large numbers and buildings with phased evacuation (see **3.68**). In a fire risk assessment, it might also be appropriate to take account of the fact that the presence of a voice alarm system reduces evacuation time.*

If it has been identified in the fire risk assessment that deaf or hard of hearing occupants are, or are likely to be, present in the building, consideration needs to be given to means for warning them in the event of fire. This might simply comprise suitable managerial arrangements, but could necessitate flashing beacons or even special means of warning, such as vibrating pagers.

*Although a facility can be provided for fire alarm signals to be transmitted automatically to an alarm receiving centre (see **3.2**) from where the fire and rescue service is summoned, this is not normally necessary for the purpose of life safety. However, there are certain occupancies in which the early summoning of the fire and rescue service is so critical, and staff levels at certain times might be so low, that there is an advantage in such a facility. Examples include certain residential care premises and hospitals.*

*Normally, the fire risk assessment considers the functionality of a fire detection and fire alarm system, but it does not involve any detailed engineering evaluation of the system. It will, however, be confirmed that the fire detection and fire alarm system is subject to routine testing and maintenance, so that faults and major shortcomings are identified by this means (see Clause **15**). Moreover, it is normally appropriate for the fire risk assessor to consider whether the fire alarm signal is likely to be audible in all relevant areas of the building, based on a visual inspection of the locations of sounders or loudspeakers, even though shortcomings are normally identified by routine testing. The fire risk assessment might then recommend, within the action plan, that an engineering evaluation, including measurement of sound pressure levels in "suspect" areas, be carried out.*

14.1.3 *Means of Escape*

*In considering the likely consequences of fire, the fire risk assessor needs to consider the likely effects of fire on escape routes (see **3.18**) during the escape time (see **3.19**), taking into account the time for detection of fire and raising the alarm (see **14.1.2**). This requires a thorough evaluation of means of escape.*

If the means of escape comply with the requirements of modern building regulations, or if the building complies with the conditions shown in a fire certificate issued within the past few years, it is unlikely that a need for major improvements will be identified in the fire risk assessment. Thus, again, suitable benchmark standards for means of escape include guidance that supports building regulations, the relevant part of BS 5588 or guidance that supports legislative requirements for fire safety in existing buildings.

However, means of escape is just one of the fire protection measures that affect the consequences of fire and, hence, the fire risk. Therefore, a departure from one or more recommendations of a prescriptive code of practice regarding means of escape might be acceptable when all other fire precautions are taken into account. Such other fire precautions include early warning of fire, rapid response to the warning by occupants and measures to increase the ASET.

*The first effect of a fire on the safety of occupants is often the presence of smoke in escape routes. This results in loss of, or reduction in, visibility. Thus, in general, adequate means of escape are provided if people can immediately, or within a short distance of travel, turn their back on any fire and move to a final exit (see **3.22**) along smoke-free escape routes.*

Four critical factors in the assessment of means of escape are therefore:

*a) the maximum distance occupants must travel to reach a place of safety (see **3.69**);*

*b) the avoidance of long dead ends (see **3.14**) in which escape is possible in only one direction;*

c) the number, distribution and widths of storey exits and final exits;

d) the means for protecting escape routes from ingress or build up of smoke that prevent occupants' escape.

*It can be anticipated that occupants with disabilities are, or are likely to be, present in the building, and consideration needs to be given to arrangements for their evacuation in the event of fire. In most multi-storey buildings, designated refuges (see **3.72**) are likely to be necessary, and there will be a need for arrangements to assist mobility-impaired occupants to escape from the refuge using staircases and/or specially designated evacuation lifts (see **3.20**). It might also be necessary for fire doors to be held open by automatic door release mechanisms (see **3.4**).*

The subject of design of means of escape is outside of the scope of this document. It is assumed that the fire risk assessor has sufficient knowledge of the principles of means of escape to assess the adequacy of the means of escape in the buildings in question. Moreover, the number of component factors that need to be considered is greater than in the case of other fire protection measures. Accordingly, Annex C sets out the key factors to consider in assessment of means of escape.

14.1.4 *Signs and notices*

In order for occupants, particularly those who are unfamiliar with the building, to use the building safely, there is normally a need to provide fire exit signs to direct people towards alternative means of escape. It is, therefore, important to consider the adequacy of such signage in the fire risk assessment.

NOTE It is a requirement of the Health and Safety (Safety Signs and Signals) Regulations 1996 that these signs incorporate the appropriate pictogram. Guidance on escape route signs is given in BS 5499-4.

In the course of the fire risk assessment, there is also a need to consider whether other forms of fire safety signs and notices are necessary, and whether those provided are adequate. Examples include:

*a) other safe condition signs (see **3.74**) (e.g. indicating use of escape hardware);*

b) signs on fire doors, indicating the need for the doors to be kept shut, kept locked or kept clear (in the case of automatically closing fire doors), as appropriate;

*c) other mandatory signs (see **3.59**), such as those indicating the need to keep a fire exit clear;*

*d) fire equipment signs (see **3.28**), particularly where, for example, fire extinguishers or hose reels are hidden from direct view;*

e) "no smoking" signs, if relevant;

f) fire procedure notices.

14.1.5 Emergency escape lighting

If escape routes require artificial illumination, there is a need to consider whether emergency escape lighting is necessary. It is not appropriate to assume that the absence of a recommendation for emergency escape lighting in the appropriate guidance documents implies that, in all circumstances, emergency escape lighting is unnecessary.

NOTE Attention is drawn to guidance documents that support building regulations (e.g. in England and Wales, Approved Document B [7]) for guidance on the need for emergency escape lighting in new buildings.

In the fire risk assessment, a judgement is necessary as to the likelihood that:

   a) *fire will cause failure of the normal lighting on any part of the escape route before all occupants have escaped from the area; and*

   b) *the loss of normal lighting will result in injury to occupants as they endeavour to evacuate the building.*

Factors to consider, therefore, are:

   1) *the length and complexity of the escape routes;*

   2) *the familiarity of the occupants with the building;*

   3) *the measures to control the development of fire;*

   4) *the measures to provide early warning of fire;*

   5) *the presence of borrowed light (e.g. from street lighting);*

   6) *the hours of work in the building;*

   7) *the presence of sleeping occupants, for whom emergency escape lighting is normally necessary;*

   8) *the presence of windowless areas.*

*If a judgement is made that emergency escape lighting is not necessary in circumstances in which it would normally be recommended in prescriptive codes of practice, it is appropriate for this to be justified in the documented fire risk assessment (see Clause **9**).*

*Normally, the fire risk assessment does not involve any detailed engineering evaluation of an emergency escape lighting system. However, it is important to confirm that the system is subject to routine testing and maintenance, so that faults and major shortcomings are identified by this means (see Clause **15**).*

*Moreover, if emergency escape lighting is considered necessary, it is normally appropriate for the fire risk assessor to consider whether the extent of an existing system is sufficient, based on a visual inspection of the areas of coverage and the provision of luminaries, and whether the duration for which emergency escape lighting can be provided is adequate. There will also be a need to consider whether maintained emergency lighting (see **3.58**) is provided where required, or whether non-maintained emergency lighting (see **3.64**) is sufficient. The fire risk assessment might, nevertheless, recommend, within the action plan, that an engineering evaluation be carried out, including verification of the adequacy of levels of illuminance. It is also normally appropriate to confirm that there are suitable facilities for routine testing of the installation.*

In many cases, an existing emergency escape lighting system will not comply in full with current recommendations (e.g. in respect of illuminance levels), particularly if the system was installed some years prior to 1999. This might be perfectly acceptable, but it is appropriate for new systems, and new work associated with upgrading of existing systems, recommended in the action plan, to comply with the current recommendations.

NOTE Attention is drawn to current recommendations in BS 5266-1 and BS 5266-7.

14.1.6 Manual fire-fighting equipment

All fires begin as small fires (other than in the case of explosions). Accordingly, it is normally appropriate for buildings to be provided with means for occupants to extinguish a fire. Normally, portable fire extinguishers are regarded as the basic provision, while hose reels tend to be regarded as optional supplementary protection.

In the fire risk assessment, consideration needs to be given to the need for manual fire-fighting equipment, the type of equipment that is necessary and the existing provision of such equipment.

NOTE The benchmark code of practice for provision of portable fire extinguishers is BS 5306-8. Guidance on the provision of hose reels is given in BS 5306-1.

*In most buildings there is the potential for class A fires (see **3.6**). Therefore, the most important fire-fighting equipment is that which is suitable for extinguishing these fires. Normally, additional extinguishers that are suitable for use on live electrical equipment (e.g. portable CO2 extinguishers) are necessary. Where there is the potential for class B fires (see **3.7**), suitable extinguishers are necessary. In kitchens, extinguishers suitable for class F fires (see **3.10**) might be necessary. It is not normal to provide extinguishers specifically for class C fires (see **3.8**), as to extinguish these fires can often result in the potential for an explosion. For certain unusual special hazards, such as class D fires (see **3.9**), appropriate extinguishers might be necessary.*

14.1.7 *Structural and similar measures to limit fire spread and development*

*In the course of the fire risk assessment, consideration needs to be given to structural and similar measures that are intended to limit the spread and development of fire within the building (in addition to consideration already given to similar measures that are specifically intended to protect means of escape). In some simple buildings in which compartmentation (see **3.12**) is not necessary for compliance with the relevant building regulations, there might be no such measures.*

*However, where compartment walls or floors are provided, some consideration needs to be given to the integrity of these. Usually, in the course of the fire risk assessment, a detailed examination of the building construction is not practicable, and there can only be visual inspection of a sample of areas (e.g. to check visually for any obvious inadequacies in fire stopping (see **3.52**)). More generally, since any structural barrier will resist the passage of smoke or fire for at least some time, obvious shortcomings in fire stopping of service penetrations need to be addressed in the action plan.*

*Traditionally, it has been regarded as good practice to enclose areas of high fire hazard in construction of appropriate fire resistance (see **3.38**). The possible need for this is, therefore, normally considered in the fire risk assessment.*

In new building work, the flammability of wall and ceiling linings is controlled under building regulations. If the linings continue to comply with the original requirements in this respect, they are likely to be satisfactory. However, consideration needs to be given to the issue of linings, as unsatisfactory linings can promote the spread and development of fire. In buildings with large areas of drapes, etc, such as cinemas and theatres, consideration normally needs to be given to their flammability. Similarly, in some buildings, the flammability of furniture and furnishings might need to be considered.

14.1.8 *Other fire protection systems*

Other fire protection systems that might need to be taken into account in the fire risk assessment include:

   a) *automatic sprinkler installations;*

   b) *smoke control systems;*

   c) *localized fire suppression systems (e.g. gaseous extinguishing systems);*

   d) *dry rising mains (see **3.15**), wet rising mains (see **3.82**) and fire-fighting lifts (see **3.32**).*

While such systems are not present in all buildings, they can have an important role to play in the safety of occupants in certain large or complex buildings. Even if the objective of such a system is property protection or assistance to the fire and rescue service, it is still appropriate to note, and take account of, the system in the fire risk assessment.

*Automatic sprinkler installations are very effective in control of fire. The presence of an automatic sprinkler installation might, therefore, enhance life safety by limiting the spread of fire from its point of origin and, in some circumstances, might allow a reduction in the performance requirements of elements of construction and compartmentation. While an engineering evaluation of an automatic sprinkler system is not normally appropriate in the course of the fire risk assessment, it is normally appropriate to confirm that there are no obvious shortcomings created by, for example, storage of stock too close to sprinkler heads. It is also appropriate to confirm that there are adequate arrangements for testing and maintenance of the system so that faults and major shortcomings are identified (see Clause **15**).*

*Similarly, in some (usually complex) buildings, smoke control systems can be essential for protection of means of escape. For example, in most shopping complexes, the combination of automatic sprinkler and smoke control systems is an essential part of the fire safety engineering package. Again, although an engineering evaluation of a smoke control system is usually outside the scope of the fire risk assessment, it is often vital to ensure that there are adequate arrangements for ongoing control, testing and maintenance of such systems (see Clause **15**).*

*Localized fire suppression systems are often provided primarily for property protection. However, they might contribute to life safety. In some cases, they might even compensate for a reduction in the standards of other fire protection measures. For example, sometimes a cooking area (e.g. within a retail floor of a shop) might require enclosure in fire resisting construction, but it might be acceptable to omit this construction if a fixed fire extinguishing system is fitted to the cooking equipment. Due account might, therefore, need to be taken of such systems, and arrangements for their testing and maintenance confirmed (see Clause **15**).*

*In most buildings that require dry or wet rising mains, or fire-fighting lifts, these will already be present. It will be unusual for a need for such facilities to first be identified in the fire risk assessment. Usually, at the stage that these facilities come into operation, the building is already evacuated, and they are primarily of assistance to the fire and rescue service. However, since safety of fire-fighters might depend on the correct operation of these facilities, arrangements for their testing and maintenance will need to be considered (see Clause **15**).*

14.2 Recommendations

The following recommendations are applicable.

a) The fire risk assessment should give consideration to, at least:

1) means for detecting fire and giving warning to occupants;
2) means of escape from the building;
3) fire safety signs and notices;

   NOTE 1 Attention is drawn to the Health and Safety (Safety Signs and Signals) Regulations 1996 [8] regarding requirements for signs relating to fire escape and fire extinguishing equipment.

4) emergency escape lighting;
5) means to limit spread and development of fire;
6) means for fighting fire;
7) other relevant fire protection systems.

The extent to which these measures are necessary, and the adequacy of existing measures, should be considered, and shortcomings in such measures should be addressed within the action plan (see Clause **18**).

NOTE 2 It is always necessary for there to be adequate means of escape in the event of fire.

NOTE 3 The fire risk assessment does not normally involve a detailed engineering evaluation of fire protection systems and equipment, but a recommendation for such an evaluation might be included in the action plan if there are doubts about the adequacy of the system.

b) The purpose of assessing the fire protection measures described in **14.2 a)** is to determine their contribution to safety of occupants in the event of fire. However, each of these measures should not be considered in total isolation of the other measures; it is appropriate to consider the effect of the entire "package" of measures on the consequences of fire to life safety.

c) In the case of fire detection and fire alarm systems, consideration should be given to the need, or otherwise, for automatic fire detection, and to the adequacy of the means for warning people in the event of fire.

d) Consideration should be given to the means for warning any deaf or hard of hearing occupants identified as especially at risk in the event of fire.

e) Consideration of means of escape should comply with the recommendations given in Annex C.

f) Consideration needs to be given to arrangements for evacuation of any sight- or mobility-impaired occupants identified as especially at risk in the event of fire.

g) In every fire risk assessment, a judgement should be made as to whether there is a need for emergency escape lighting. If emergency escape lighting is considered necessary, subjective consideration should be given to the adequacy of any existing emergency escape lighting.

h) Consideration should be given to the adequacy of the type, number and sitting of manual fire-fighting appliances.

   NOTE 4 Normally, a sufficient number of portable fire extinguishers that are suitable for use on class A fires should be provided.

i) Consideration should be given to the adequacy of fire stopping, the flammability of linings and, where appropriate, the flammability of furniture and furnishings.

NOTE 5 It is not normally practicable to carry out a complete review of fire stopping in a building, and reliance on a visual inspection of a sample of readily accessible areas will normally be adequate.

j) The fire risk assessment should take account of all other fire protection systems, including automatic sprinkler systems, smoke control systems, localized fire suppression systems, dry or wet rising mains and fire-fighting lifts. Even if the objective of such systems or facilities is considered to be property protection, their presence should be noted in the fire risk assessment, and due account should be taken of their contribution (if any) to safety of occupants from fire.

15 Assessment of fire safety management

15.1 Commentary

15.1.1 *General*

In the fire risk assessment, fire safety management (see **3.45**) *needs to be regarded as of equal importance to fire protection measures. In its broadest sense, fire safety management includes certain policies and procedures designed to prevent the occurrence of fire by eliminating or controlling fire hazards. However, most of these aspects of fire safety management have already been considered in Clause* **12**.

Fire safety management also includes the following:

   a) *designated responsibility for fire safety in the building;*

   b) *access to suitable advice on the requirements of fire safety legislation;*

   c) *procedures for people to follow in the event of fire, including people with special responsibilities;*

   d) *nomination of people to respond to fire and, where appropriate, to assist with evacuation;*

   e) *arrangements for liaison with the fire and rescue service, both in respect of pre-planning for fire and at the time of any fire;*

   f) *arrangements for routine inspections of the building and its fire precautions or for more formal fire audits (see* **3.23**);

   g) *staff training and fire drills;*

   h) *testing and maintenance of fire protection systems and equipment;*

   i) *keeping appropriate records;*

   j) *implementation of the action plan's recommendations;*

   k) *review of the fire risk assessment at appropriate intervals.*

Points a) to i) are discussed in the sub-clauses that follow. Point j) is discussed in Clause **18** *and k) is discussed in Clause* **19**.

15.1.2 *Responsibility for fire safety*

Although legislation does not demand that a specific, named person be responsible for fire safety within a particular building, it is of advantage to confirm, in the fire risk assessment, that, within the organization, there is someone who, in at least an administrative sense, is responsible for fire safety within the building. The intention is not to provide a legal interpretation of responsibility, but to reflect the managerial arrangements in place at the time of the fire risk assessment. The responsible person might, or might not, have a legal responsibility for breaches of legislation, etc.

According to the manner in which the organization is structured, the person named in this section of the fire risk assessment might be a director, building manager, facilities manager, health and safety manager, fire safety manager, estates manager, etc. The person might, or might not, work within the building, and the responsibility could even be shared by two or more people. It is, however, important in the management of any organization that someone is, and accepts that they are, responsible for fire safety, particularly in the case of a building in multiple occupation.

15.1.3 Access to advice

NOTE Attention is drawn to the Management of Health and Safety at Work Regulations for requirements on the definition and appointment of a "competent person" to assist in compliance the Fire Precautions (Workplace) Regulations.

The "competent person" required by the Management of Health and Safety at Work Regulations might, or might not, be the person responsible for fire safety, to which **15.1.2** makes reference. However, the two will often be different, since the person having responsibility for fire safety might be a senior manager, while the "competent person" might be a trained professional in the field of fire safety or health and safety.

Given this situation, it is appropriate to consider in the fire risk assessment the arrangements for assistance to an organization in compliance with the Workplace Fire Precautions Legislation and in devising and applying suitable fire precautions. This does not imply that external consultants must be appointed. Often, organizations are able to appoint one or more of their own employees for this purpose, while large organizations might appoint whole departments with specific health and safety responsibilities, including specialists in various matters, such as fire safety. Equally, if consultants are used for advice, it is necessary for their activities to be co-ordinated by the organization, since external consultants will usually be appointed in an advisory capacity only, and their appointment does not absolve the organization from its responsibilities (see Clause **6**).

15.1.4 Fire procedures

In the course of the fire risk assessment, there is a need to ensure that there are formal, documented procedures for people to follow in the event of fire, and that the procedures in question are adequate. Adequate procedures will address:

   a) actions to follow on discovery of fire;

   b) actions to follow on hearing the fire alarm signal;

   c) the importance of raising the alarm immediately on discovery of fire;

   d) the importance of evacuating the building immediately when the fire alarm sounds;

   e) the arrangements for evacuation of disabled occupants;

   f) the policy on whether employees should, or should not, attempt to tackle a fire;

   g) the summoning of the fire and rescue service;

   h) the location of evacuation assembly points;

   i) the importance of not attempting to re-occupy the building until instructed to do so by the fire and rescue service.

   NOTE In cases of false alarms, where the fire and rescue service does not attend the building, the decision to re-enter the building will need to be taken by a responsible person.

Normally, there will be a need for special procedures for occupants with special duties in the event of fire. These could include, for example:

   1) switchboard operators (in relation to, for example, summoning of the fire and rescue service);

   2) fire wardens (see **3.53**);

   3) assembly point wardens;

   4) those responsible for meeting the fire and rescue service;

   5) security personnel;

   6) senior management.

15.1.5 Nomination of people with special duties in the event of fire

In carrying out the fire risk assessment, there is a need to ensure that there are adequate arrangements for summoning the fire and rescue service in the event of fire. The arrangements will form part of the fire procedures for the building (see **15.1.4**), but it might be the case that summoning of the fire and rescue service is the responsibility of a nominated post-holder, such as a switchboard operator.

The fire risk assessor needs to investigate the arrangements in place for fire-fighting and to ensure that these are adequate. For example, the fire procedures might dictate that anyone who discovers a fire may *tackle the fire with fire extinguishing appliances if it is safe to do so; alternatively, only a proportion of the staff in the building might be authorized to do so.*

NOTE 1 Attention is drawn to the Fire Precautions (Workplace) Regulations' requirements for employers regarding arrangements for fire-fighting.

The fire risk assessor also needs to investigate the arrangements for ensuring that the building is evacuated (e.g. by appointment of fire wardens), and to ensure there is proper control, co-ordination and monitoring of evacuation procedures. Information on the status of the evacuation will be of importance to the fire and rescue service when they arrive at the building.

NOTE 2 Attention is drawn to the Management of Health and Safety at Work Regulations' requirements regarding the nomination of people to assist in evacuation.

In buildings such as residential care homes, it is appropriate to consider, within the fire risk assessment, whether sufficient levels of staff are present to ensure the safety of residents during both day and night. This will normally necessitate discussions with the building management.

NOTE 3 Attention is drawn to Local Authority registration conditions relating to staffing levels.

15.1.6 Liaison with the fire and rescue service

In large and complex buildings, it is important that there are arrangements for local fire and rescue service crews to familiarize themselves with the building and, with, for example, the facilities for fire-fighting and potential risks to fire-fighters. In some such buildings, there might be a need for pre-planning emergency procedures with the fire and rescue service. In addition, it is important that the fire procedures for the building include arrangements for summoning of the fire and rescue service in the event of fire and meeting the fire and rescue service on arrival.

15.1.7 Routine inspections

The fire risk assessment is somewhat similar to the MOT inspection of a car; it reflects the conditions found by an assessor at a particular point in time. There is, however, a need to ensure that, on a more routine basis, there are means for detecting deficiencies in fire precautions. Accordingly, it is appropriate for the fire risk assessor to investigate arrangements for suitably trained or instructed building occupants to carry out routine inspections of the fire precautions.

Such inspections need no specialist knowledge, but can make a major contribution towards the maintenance of adequate fire precautions by checking that, for example, manual call points, fire detectors, sprinkler heads, etc. remain unobstructed, self-closing fire doors operate correctly, fire exit doors that are not in normal use open easily and that there is no storage in escape routes that should remain relatively sterile (e.g. protected staircases). Sometimes these matters are addressed in the course of health and safety inspections or more specific fire audits. Often, more frequent day-to-day inspections, of a basic nature, can be carried out by, for example, patrolling security officers.

15.1.8 Staff training and fire drills

*Since failure of people to react correctly has been associated with many fires that have resulted in serious loss of life, an important part of the fire risk assessment is consideration of arrangements for giving instruction and training to staff on fire safety matters and for carrying out fire drills. Fire safety induction training (see **3.44**) for all new staff is particularly important.*

NOTE Attention is drawn to the Management of Health and Safety at Work Regulations regarding adequate training for employees.

*Thereafter, fire safety refresher training (see **3.50**) needs to be given periodically. The frequency of refresher training needs to take into account the turnover of staff, the complexity of the building and the fire procedures, and the fire risk. There will often be a need to provide additional, or special, training for people who have special responsibilities in the event of fire; this could, for example, include fire wardens.*

Legislation does not always specifically require that fire drills are carried out. However, generally fire drills are important in all except the smallest building. The drills are a means of reinforcing training, and provide feedback on the effectiveness of the training that has been carried out.

15.1.9 Testing and maintenance of fire protection measures

The fire risk assessor needs to ensure that there are adequate arrangements for testing and maintenance of all fire protection measures. There is also a need to ensure that the workplace itself is adequately maintained in order to avoid certain fire hazards.

NOTE Recommendations for testing and maintenance of systems are given in the relevant British Standards for the particular systems and equipment.

15.1.10 *Record keeping*

Legislation does not necessarily specifically require that records of training, inspection, testing, maintenance, etc. are kept. Nevertheless, such records are an important means of demonstrating, if required, that all legislative obligations have been satisfied. It is, therefore, relevant for the fire risk assessor to consider any records that exist and to make recommendations, where appropriate, for keeping of records. These records can also be important in demonstrating that there have been no breaches of good practice that could result in litigation in the event of injury to an occupant of the building in the event of fire.

*In addition, there will be a need for a well-documented fire safety manual (see **3.47**) in large and complex buildings.*

15.2 Recommendations

The following recommendations are applicable.

a) The fire risk assessment should record the name(s) or post(s) of the person(s) responsible for fire safety in the building.

b) It should be confirmed that there are arrangements for obtaining competent advice on the requirements of fire safety legislation. The source of such advice should be recorded in the documented fire risk assessment (see Clause **9**).

c) In the course of the fire risk assessment, the following matters should be considered. Any shortcomings in these matters should be identified in the documented fire risk assessment and should be addressed in the action plan (see Clause **18**).

   1) The fire procedures, including procedures for people with special responsibilities in the event of fire.
   2) The arrangements for summoning the fire and rescue service in the event of fire.
   3) The nomination of people to respond to fire, using fire-fighting equipment if appropriate to do so.
   4) Where appropriate, the nomination of people to assist with evacuation.
   5) Arrangements for liaison with the fire and rescue service.
   6) Arrangements for routine inspections of the building and its fire precautions.
   7) Staff training.
   8) Fire drills.
   9) Testing and maintenance of fire protection systems and equipment.
   10) Maintenance of the workplace.
   11) Appropriate records, including, where applicable, a fire safety manual.

16 Assessment of likely consequences of fire

16.1 Commentary

*Once all fire protection measures and all aspects of fire safety management have been assessed, the fire risk assessor is in a position to make an assessment of the likely consequences of fire, taking into account the factors concerning the building and its occupants discussed in Clause **11**. As well as consideration of fire protection measures and matters such as fire procedures, account needs to be taken of human behaviour. It is not, for example, appropriate for the fire risk assessment to assume error-free perfection in the response of people to fire alarm signals. Consideration needs to be given to the manner in which the known occupants of the building are likely to behave in the event of fire.*

*It would be possible, in theory, to associate different consequences of fire with different fire scenarios arising from each of the fire hazards identified in the fire hazard identification step of the fire risk assessment (see Clause **12**). However, this would make the fire risk assessment process unnecessarily complex and unduly lengthy. Usually, it is sufficient to consider the most likely consequences of a fire in the building, taking into account the range of fire scenarios that it is reasonable to anticipate, and assuming that only one fire occurs at any one time (i.e. generally discounting multiple seats of fire).*

Consequences need to take into account the extent of injury that would occur to occupants in anticipated scenarios, and take into account the number of occupants likely to be affected. Consequences are more serious if a greater number of occupants are affected. Equally, serious consequences include, for example, a situation in which there is a high likelihood that a small number of occupants (even one) will be subject to serious injury in the event of fire.

The likely consequences of fire need not, and usually cannot, be expressed in a statistical manner (e.g. probability of death or serious injury). All that is required is a subjective judgement that classifies likely consequences of fire into one of several pre-determined categories. Since the assessment of these factors is subjective, the use of numbers to express the likely consequences of fire does not confer any greater accuracy to the assessment of fire risk.

The pre-determined categories of likely consequences of fire may be described in the form of words, such as "slight harm", "moderate harm" and "extreme harm", provided these terms are defined, or in the form of numbers (e.g. 1, 2 and 3), but there will be a need for at least three categories. However, if likely consequences are expressed in the form of numbers, care is necessary to ensure that it is not implied, for instance, that a score for likely consequences of "2" indicates that fire is twice as likely to result in casualties compared to a score of "1".

There is no upper limit to the number of categories of likely consequences that can be adopted in the fire risk assessment process. However, if too many categories are adopted, the distinctions between categories will be meaningless. Moreover, if the same fire risk assessment process is then applied to numerous different buildings (e.g. within the estate of a single organization), particularly by different fire risk assessors, assessments of the likely consequences of fire are likely to be inconsistent, and the benefits of comparing fire risk in different buildings (e.g. for the purpose of prioritizing improvements on a building-by-building basis) will then be lost.

*It is common practice for the assessment of fire risk (see Clause **17**) to be expressed by means of combining the assessment of the likelihood of fire and the assessment of the likely consequences of fire, using a matrix. It is this method that is adopted in this PAS. Where such an approach is adopted, it is helpful to use the same number of categories for both likelihood of fire and likely consequences of fire.*

*In assessing the likely consequences of fire, for the purpose of carrying out the fire risk assessment to which this PAS relates, it is not normally necessary, or appropriate, to carry out calculations of the type used in the practice of fire safety engineering (see **3.43**). However, the principles of fire safety engineering may be used, in a subjective manner, to assess the likely consequences of fire, using the principle of "timelines" that forms the basis of fire safety engineering (see Figure 3).*

*In Figure 3, which is reproduced (with minor modifications) from BS 7974, the escape time (see **3.19**) is broken down into a number of components, namely:*

- *a) the time between ignition of a fire and detection of the fire (whether by occupants or by an automatic fire detection system);*
- *b) the time between detection and the giving of the alarm warning to occupants;*
- *c) the time between the giving of the alarm warning and the recognition by occupants that the alarm warning is a warning of fire;*
- *d) the time between this recognition and response by occupants (i.e. the time to begin evacuation);*
- *e) the time between response and completion of evacuation of occupants to a place of safety.*

*The escape time, so derived, is then compared with the ASET (see **3.5**). For safe evacuation of occupants, the ASET must be longer than the escape time. In the fire risk assessment, Figure 3 is particularly useful in forming the basis for an analytical approach to situations in which fire protection measures (such as means of escape) do not conform to the recommendations of the relevant prescriptive code of practice.*

For example, if the travel distances are significantly longer than prescribed in the relevant code of practice, so extending the travel time and hence the escape time, account may be taken of fire precautions (whether existing or proposed in the action plan) that act to reduce the escape time by a commensurate amount; such fire precautions may be other fire protection measures or various enhancements in fire safety management (e.g. fire procedures, fire training and fire drills). Alternatively, account may be taken of fire precautions (whether existing or proposed in the action plan) that extend the ASET (e.g. measures to control smoke).

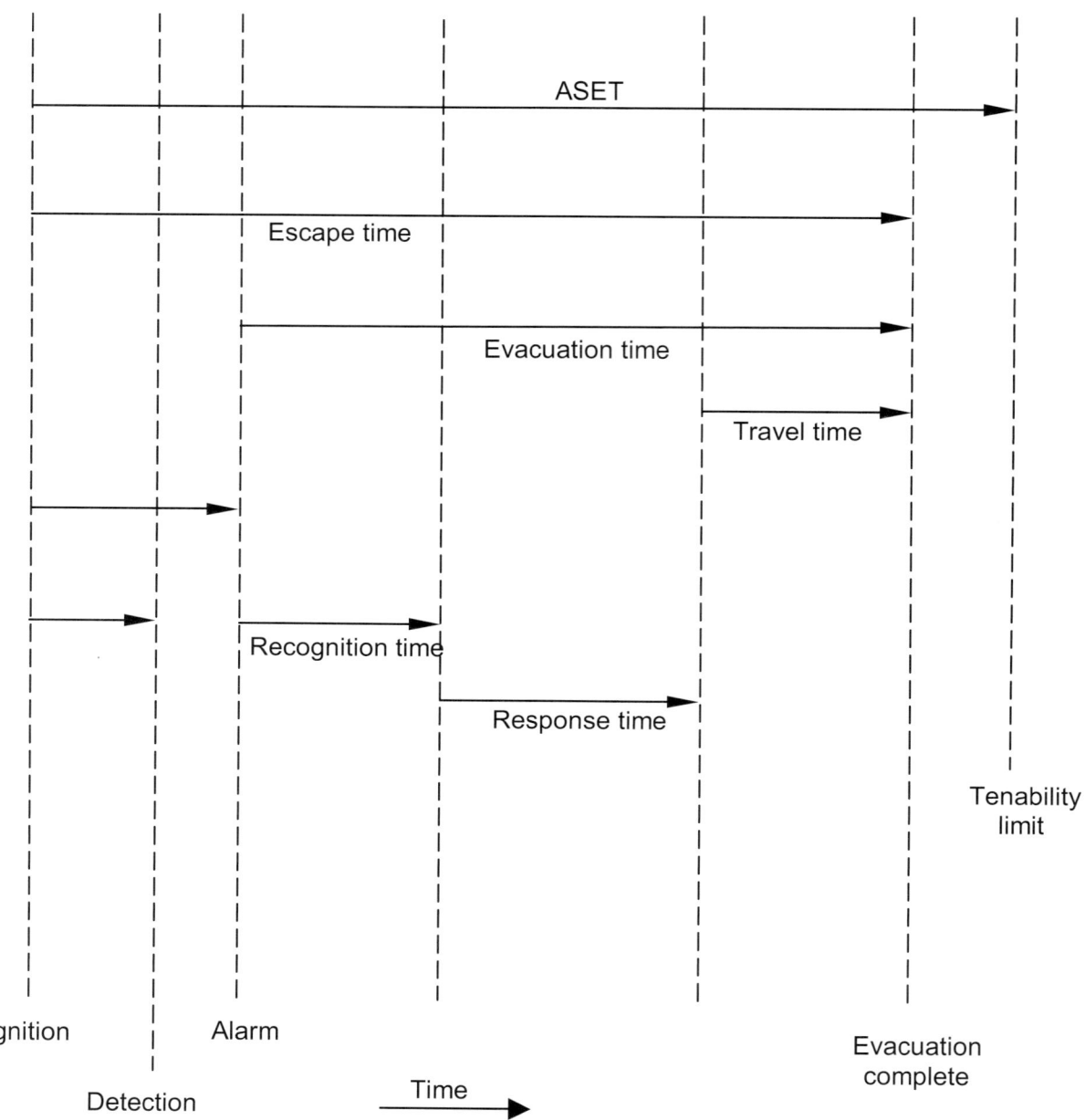

Figure 3 — Example of timeline comparison between ASET and escape time (reproduced from BS 7974)

16.2 Recommendations

The following recommendations are applicable.

a) In the process of every fire risk assessment, an assessment should be made of the likely consequences of fire. It is usual and acceptable for the likely consequences of fire to be expressed subjectively (e.g. "slight harm", "moderate harm" or "extreme harm").

b) If, in the fire risk assessment methodology adopted, a matrix is used to combine likelihood of fire and likely consequences of fire in order to determine the fire risk, the number of pre-determined categories of likely consequences of fire should be the same as the number of pre-determined categories of likelihood of fire (see **13.2 b)**).

17 Assessment of fire risk

17.1 Commentary

It is innate to the process of carrying out the fire risk assessment that there be an assessment of fire risk, which it is then appropriate to document. The assessment of fire risk enables the (usually subjectively based) fire risk in one building to be compared with the fire risk in other buildings (e.g. within the single estate of one organization), so identifying those buildings in greatest need of attention. Even applied to a single building in isolation, the assessment of fire risk can provide a useful descriptor that imparts a sense of the magnitude of fire risk.

The categories for classification of fire risk are derived from the those used to determine the likelihood and likely consequences of fire (see Clauses 13 and 16). Whereas it is normally sufficient to classify likelihood of fire, or likely consequences of fire, into one of three pre-determined categories, a greater number of categories of fire risk is normally appropriate in order to cater for the range of levels of fire risk that can occur. Thus, a minimum of five pre-determined categories of fire risk is normally appropriate.

The category of fire risk for any building may be determined by combination of the likelihood of fire and the likely consequences of fire, using a matrix; this is a method of risk assessment commonly adopted in the field of health and safety. The matrix given in Table 1 is based on one given in BS 8800 and may be adopted in assessment of fire risk.

The advantage of this approach is that it tends to result in relatively consistent assessments of risk (and, hence, fire risk) by different risk assessors; the risk assessor need "plug in" to the matrix only one of three levels of likelihood and one of three levels of likely consequences, but can derive thereby any one of five levels of (fire) risk.

Table 1 — A simple risk level estimator.

	Slight harm	Moderate harm	Extreme harm
Highly unlikely	TRIVIAL RISK	TOLERABLE RISK	MODERATE RISK
Unlikely	TOLERABLE RISK	MODERATE RISK	SUBSTANTIAL RISK
Likely	MODERATE RISK	SUBSTANTIAL RISK	INTOLERABLE RISK

17.2 Recommendations

The following recommendations are applicable.

   a) In the process of every fire risk assessment, an assessment should be made of the fire risk in the building. It is usual and acceptable for the fire risk to be expressed in terms of one of a number of pre-determined categories of risk (e.g. "trivial", "tolerable", "moderate", "substantial" or "intolerable").

   b) If, in the fire risk assessment methodology adopted, fire risk is expressed in terms of one of several pre-determined categories, the number of pre-determined categories should be at least five.

   c) The fire risk assessment methodology adopted should be such that there is a transparent means for combining the likelihood of fire and the likely consequences of fire to derive the fire risk (e.g. use of a matrix with pre-determined categories for each of these).

18 Formulation of an action plan

18.1 Commentary

*The outcome, and indeed the principal raison d'être, of the fire risk assessment is the action plan. The action plan comprises recommendations that are intended to ensure that the fire risk is reduced to, or maintained at, a tolerable level (see **3.79**). Even if the fire risk is already tolerable, there is often a need to make recommendations in the action plan, often involving low cost or changes in managerial arrangements, to address minor deficiencies in fire precautions.*

In formulating an action plan for buildings in which the fire risk has been assessed as unacceptably high, the analytical approach to fire risk assessment permits backtracking to determine whether, in effect, the problem arises from inadequate fire prevention (i.e. inadequate means for control or elimination of fire hazards), inadequate fire protection (e.g. unsatisfactory means of escape or fire warning systems), shortcomings in fire safety management, or a combination of these.

The action plan is an inventory of actions, often prioritized, to devise, maintain or improve controls. Ideally, the inventory will include measures to eliminate or control hazards (e.g. better separation of combustible materials and ignition sources). A blend of physical and procedural controls is often necessary.

The adequacy of the action plan needs to be tested, at least in the mind of the fire risk assessor, before it is finalized. At that stage, it is appropriate to consider the following questions.

   a) Will the revised controls lead to tolerable fire risk levels?

   b) Are new hazards created?

   c) Have the most cost-effective solutions been chosen?

   d) What will occupants affected think about the need for, and practicality of, the revised fire precautions?

   e) Will the revised fire precautions be adopted and maintained in practice and not ignored in the face of, for example, normal use of, and operations in, the building?

*All of these questions have a relevance to any action plan, the objective of which must, of course, be to achieve tolerable risk, but without the creation of new hazards. The fire precautions proposed ought to be the most cost effective available; often a single fire safety objective (see **3.48**) can be satisfied by a variety of measures.*

The practicality of fire precautions, and their acceptability to occupants, are also essential. There is no point in installing self-closing fire doors if discussion with occupants would have revealed that they would be such an impediment to the work process that they would always be wedged in the open position. Equally, if this is clear from discussion with those in the workplace, the problem may be pre-empted by installing fire doors that are held open by automatic door release mechanisms, which release the self-closing doors on operation of the fire alarm system.

Often, it is appropriate to allocate priorities to each measure recommended in the action plan, to reflect the urgency of the measure, as determined in the fire risk assessment. (This might, however, be unnecessary if, for example, most of the recommended measures are minor in nature and will be implemented in the short term in any case.)

If prioritization is appropriate, a scheme of prioritization that is suitable for the way in which the company operates and projects are planned is often helpful. There is no right or wrong scheme of prioritization, but, whatever scheme is adopted, it needs to be simple to understand, consistent to apply and relatively unambiguous as far as allocation of priorities is concerned. This suggests that it is appropriate for there to be no more than three or four priorities.

A simple scheme might be one with only three priorities, such as:

　　1) *immediate (should be implemented as soon as possible);*

　　2) *short term (should be implemented within, say, three months);*

　　3) *long term (should be implemented as and when the opportunity arises).*

Many other systems of prioritization are possible. For example, priorities might distinguish between matters that constitute breaches of legislation and those that do not.

NOTE *Under the Workplace Fire Precautions Legislation, a breach of the Regulations constitutes a criminal offence if, inter alia, the breach results in the risk of serious injury or death of one or more employees in the event of fire.*

Thus, for example, a possible scheme of prioritization could be:

　　i) *serious breach of legislation, having the potential for serious injury to occupants;*

　　ii) *matters that breach legislation but are not considered to constitute a serious threat to life safety;*

　　iii) *matters that should be addressed as good practice, but that do not constitute any significant threat to occupants.*

The implications, in terms of timescales, etc. would naturally flow from this.

Yet another possible scheme could take into account both the cost benefit and the practicality of implementation. For example, minor housekeeping items could be regarded as suitable for immediate implementation, simply because there is no reason to delay doing so, regardless of whether there is a major benefit to the safety of occupants. However, matters that might address a greater threat to occupants might be impossible to implement immediately, in the literal sense of the term, simply because specifications need to be drawn up, tenders obtained, etc.

18.2 Recommendations

The following recommendations are applicable.

　　a) Every documented fire risk assessment should incorporate an action plan. If the fire risk and existing fire precautions are such that no recommendations for improvements are necessary, it should be explicit within the document that, in the opinion of the fire risk assessor, the only actions necessary are those to maintain the existing standard of fire precautions.

　　　NOTE The action plan is sometimes, more simply, described as "recommendations", particularly when the fire risk assessment is carried out by a third party fire risk assessor (see **3.78**).

　　b) The action plan should be such as to ensure that, if implemented, it will reduce fire risk to, or maintain fire risk at, a tolerable level.

　　c) Where appropriate, the action plan should address both physical fire precautions and managerial issues.

　　d) The action plan should be both practicable to implement and possible to maintain, taking into account the nature of the building, its occupants and the work processes carried out.

　　e) The measures recommended in the action plan should be cost-effective in reducing fire risk.

　　f) No new significant hazards should result from implementation of the action plan.

　　g) The action plan should contain information regarding the appropriate effort and urgency associated with the measures recommended. Effort and urgency should be proportionate to fire risk, but financial considerations should also be considered.

19 Periodic review of fire risk assessments

19.1 Commentary

*The documented fire risk assessment is not intended to be a fire safety manual (see **3.47**), albeit that such a manual is a valuable asset in the management of fire safety, particularly in large or complex buildings. However, the fire risk assessment is a living document, in that it cannot remain valid for an unlimited length of time.*

The fire risk assessment is likely to cease to be valid when, for example:

   a) *a material alteration (see **3.61**) takes place;*

   b) *a significant change occurs in the "given" factors that were taken into account when the fire risk assessment was carried out (see **5.1**);*

   c) *a significant change in fire precautions occurs.*

Significant changes in the "given" factors could, for example, comprise a large increase in the number of occupants of the building, use of the building by significantly more disabled occupants, or introduction of a much more hazardous process. Significant changes in fire precautions include major changes in the provision or design of fire protection measures and major changes in the measures for control or elimination of fire hazards, but also include changes resulting from more gradual deterioration of fire precautions as a result of constant use or lack of maintenance (e.g. wear and tear on fire doors). Gradual changes can also occur as a result of changes in management, turnover of employees and minor changes in layout that, after a prolonged period and numerous changes, have a significant effect on means of escape.

It follows, therefore, that when any of the acute step changes described above occur, the fire risk assessment needs to be reviewed. There might also be need for approval of such changes under building regulations and/or by the fire authority. However, as gradual changes over a long period of time can also affect the validity of the fire risk assessment, there is a need for periodic review of the fire risk assessment, even if there are no obvious changes that affect its validity. In fire risk assessments carried out in accordance with this PAS, judgement of this maximum period after which the fire risk assessment needs to be reviewed, on a routine basis, is actually part of the fire risk assessment process.

When the fire risk assessment is reviewed, consideration needs to be given to the extent to which the original action plan has been implemented. Work that has not been completed needs to be identified.

*There is no correct or incorrect frequency for the routine review of the fire risk assessment. This is a matter for the judgement of the fire risk assessor and, to some extent, the organization's own fire safety policy (see **3.49**). It is, however, appropriate to take account of the likely frequency of significant changes.*

For example, the fire risk assessment for a retail outlet, in which significant changes in sales layout are likely to occur frequently, might need more frequent review than the fire risk assessment for a barrister's chambers that have remained unaltered for many decades. Also, if, at the time of the fire risk assessment, there are major shortcomings in fire precautions, the action plan will contain proposals for significant changes. These changes are likely to take place within a relatively short time, after which review of the fire risk assessment might be warranted.

Review of the fire risk assessment is not synonymous with a new assessment. Equally, however, in a routine periodic review, all aspects of the original fire risk assessment might need to be revisited to ensure that they have not been subject to change; this emphasizes the importance of adequate recording of the significant findings of the original fire risk assessment, so that the basis for its conclusions can be readily re-examined. On the other hand, if the review has arisen purely as the result of a specific material alteration, it might be the case that a much more limited review is sufficient.

*Annex D contains a pro-forma that is considered a suitable and sufficient means for documenting a periodic review of an existing fire risk assessment. The pro-forma contained in Annex D is only a model, in that, if completed by a competent person (see Clause **7**), the scope of the documented review of the fire risk assessment will comply with the recommendations of this PAS. Equally, the format of a documented review may vary from that shown in Annex D, provided that the recommendations of this clause are satisfied.*

The original fire risk assessment, in conjunction with one or more documented periodic reviews, constitutes a form of audit trail that demonstrates ongoing control of fire safety. After a period of time in which there have, for example, been several periodic reviews in which significant changes and the need for new risk control measures have been identified, the audit trail is likely to become unwieldy. At that stage, the documentation of a new and complete fire risk assessment might be appropriate.

19.2 Recommendations

The following recommendations are applicable.

a) The fire risk assessment should be subject to review when:

   1) material alterations to the building take place;
   2) a significant change occurs in the matters taken into account when the fire risk assessment was carried out;
   3) a significant change in fire precautions occurs;
   4) there is any other reason to suspect that the original fire risk assessment might no longer be valid;
   5) a defined period of time, which should be recorded in the original fire risk assessment (see **10.2 a) 9)**), has elapsed.

b) When the fire risk assessment is reviewed, it should be confirmed whether work recommended in the original action plan has been carried out correctly.

c) The fire risk assessment review frequency should take into account the likely frequency of significant alterations to the building, and should also take account of the period after which major changes in fire precautions are likely to have taken place as a result of the measures recommended in the action plan.

d) The fire risk assessment review should explicitly address the issues considered in the original fire risk assessment, albeit that less detail in the record of the significant findings is necessary, particularly in respect of fire precautions that have not changed since the original fire risk assessment.

e) The fire risk assessment review should record the name of the fire risk assessor(s), the date(s) on which the periodic review was carried out and the name(s) of the principal person(s) with whom there was consultation (e.g. for supply of relevant information) at the time of the periodic review.

f) The fire risk assessment review should record the date by which the next periodic review should be carried out.

NOTE Use of the pro-forma in Annex D enables compliance with the recommendations of this sub-clause, but, equally, other formats may be used, provided that, as a minimum, the information and matters included in the pro-forma in Annex D are addressed.

Annex A (informative)
Model pro-forma for documentation of the fire risk assessment

A.1 This Annex contains a model pro-forma for documentation of the fire risk assessment. If the pro-forma is completed by a competent person, the format and scope of the fire risk assessment will be suitable and sufficient to satisfy the recommendations of this PAS.

A.2 The format of a documented fire risk assessment may vary from that shown in this Annex, provided the recommendations of each clause of this PAS are satisfied. For example, in the case of means of escape, compliance with Annex C necessitates that the key factors in Table C.1 are explicitly addressed in the documented fire risk assessment, but not all the specific issues shown in Table C.1 and in the pro-forma contained in this Annex need necessarily be included in all documented fire risk assessments complying with the recommendations of this PAS.

A.3 Equally, the prompt list of fire hazards shown in the pro-forma may be expanded. This may be appropriate, for example, if there are significant fire hazards for which no headings are included in the pro-forma.

A.4 Where description of any fire hazards or fire precautions is considered appropriate, this can be recorded under the relevant "Comments" heading in the pro-forma. The comments sections can also be used to set out justification for acceptance of standards of any fire protection measures that depart significantly from a prescriptive norm (see **9.2 b)**).

A.5 While it might not be essential to record further information in every comments section, care needs to be taken to ensure that the pro-forma does not become a mere ticklist with inadequate supporting information.

PAS 79:2005

WORKPLACE FIRE PRECAUTIONS LEGISLATION

FIRE RISK ASSESSMENT

Employer or other Responsible Person:

Address of Property:

Person(s) Consulted:

Assessor:

Date of Fire Risk Assessment:

Date of Previous Fire Risk Assessment:

Suggested Date for Review[1] :

The purpose of this report is to provide an assessment of the risk to life from fire in these buildings, and, where appropriate, to make recommendations to ensure compliance with fire safety legislation. The report does not address the risk to property or business continuity from fire.

[Date]

---

[1] This fire risk assessment should be reviewed by a competent person by the date indicated above or at such earlier time as there is reason to suspect that it is no longer valid or there have been significant changes.

GENERAL INFORMATION

1. THE BUILDING

1.1 Number of floors:

1.2 Approximate floor area: m_ per floor.
                              m_ gross.
                              m_ on ground floor.
                              [*delete units as appropriate*]

1.3 Brief details of construction:

1.4 Occupancy:

2. THE OCCUPANTS

2.1 Approximate maximum number:

2.2 Approximate maximum number of employees at any one time:

2.3 Maximum number of members of public:

3. OCCUPANTS AT SPECIAL RISK

3.1 Sleeping occupants:

3.2 Disabled occupants:

3.3 Occupants in remote areas:

3.4 Others:

4. FIRE LOSS EXPERIENCE

5. OTHER RELEVANT INFORMATION

6. RELEVANT FIRE SAFETY LEGISLATION

6.1 The Fire Precautions (Workplace) Regulations 1997 (as amended) apply to this building: ☐

6.2 The Fire Precautions Act 1971 applies to this building, but a fire certificate **is not** required because the building:

- has been exempted from a fire certificate by the fire authority. ☐

- is put to a designated use, but a fire certificate is not required under the terms of the Designation Order. ☐

6.3 The Fire Precautions Act 1971 applies to this building, and a fire certificate is required by virtue of:

- the number/location of guests and staff sleeping in the building. ☐

- the number/location of people employed at any one time. ☐

- the use of highly flammable or explosive materials. ☐

6.4 A fire certificate has been issued:

- under the Fire Precautions Act 1971. ☐

- under the Offices, Shops and Railway Premises Act 1963. ☐

- under the Factories Act 1961. ☐

- under the Fire Certificates (Specific Premises) Regulations 1976. ☐

Certificate number: _____   Date of issue: _____

6.5 An application for a fire certificate was made on: _____

A notice specifying steps to be taken has not yet been issued by the fire authority: ☐

A notice specifying steps to be taken was issued on: _____

and requires that work is completed by: _____

6.6 An application for a fire certificate should be made as soon as possible. ☐

6.7 Other relevant fire safety legislation:

6.8 Comments:

FIRE HAZARDS AND THEIR ELIMINATION OR CONTROL

7. ELECTRICAL SOURCES OF IGNITION

7.1 Reasonable measures taken to prevent fires of electrical origin?   Yes ☐   No ☐

7.2 More specifically:

     Fixed installation periodically inspected and tested?   Yes ☐   No ☐

     Portable appliance testing carried out?   Yes ☐   No ☐

     Suitable policy regarding the use of personal electrical appliances?   Yes ☐   No ☐

     Suitable limitation of trailing leads and adapters?   Yes ☐   No ☐

7.3 Comments and hazards observed:

8. SMOKING

8.1 Reasonable measures taken to prevent fires as a result of smoking?   Yes ☐   No ☐

8.2 More specifically:

     Smoking prohibited in the building?   Yes ☐   No ☐

     Smoking prohibited in appropriate areas?   Yes ☐   No ☐

     Suitable arrangements for those who wish to smoke?   Yes ☐   No ☐

     Absence of any evidence of breaches of policy?   Yes ☐   No ☐

8.3 Comments and hazards observed:

9. ARSON

9.1 Does basic security against arson by outsiders appear reasonable[2]?   Yes ☐   No ☐

9.2 Is there an absence of unnecessary fire load in close proximity to the building or available for ignition by outsiders?   Yes ☐   No ☐

9.3 Comments and hazards observed:

---

[2] Note: Reasonable only in the context of this fire risk assessment. If specific advice on security (including security against arson) is required, the advice of a security specialist should be obtained.

PAS 79:2005

10. PORTABLE HEATERS AND HEATING INSTALLATIONS

10.1 Is the use of portable heaters avoided as far as practicable?   Yes ☐   No ☐

10.2 If portable heaters are used,

     is the use of the more hazardous type (e.g. radiant bar fires or lpg appliances) avoided?   N/A ☐   Yes ☐   No ☐

     are suitable measures taken to minimize the hazard of ignition of combustible materials?   N/A ☐   Yes ☐   No ☐

10.3 Are fixed heating installations subject to regular maintenance?   N/A ☐   Yes ☐   No ☐

10.4 Comments and hazards observed:

11. COOKING

11.1 Are reasonable measures taken to prevent fires as a result of cooking?   N/A ☐   Yes ☐   No ☐

11.2 More specifically:

     Filters changed and ductwork cleaned regularly?   N/A ☐   Yes ☐   No ☐

     Suitable extinguishing appliances available?   Yes ☐   No ☐

11.3 Comments and hazards observed:

12. LIGHTNING

12.1 Does the building have a lightning protection system?   Yes ☐   No ☐

12.2 Comments and deficiencies observed:

13. OTHER SIGNIFICANT IGNITION SOURCES THAT WARRANT CONSIDERATION

13.1 Ignition sources:

13.2 Comments and deficiencies observed:

14. HOUSEKEEPING

14.1 Is the standard of housekeeping adequate? Yes ☐ No ☐

14.2 More specifically:

Combustible materials appear to be separated from ignition sources? Yes ☐ No ☐

Avoidance of unnecessary accumulation of combustible materials or waste? Yes ☐ No ☐

Appropriate storage of hazardous materials? N/A ☐ Yes ☐ No ☐

Avoidance of inappropriate storage of combustible materials? Yes ☐ No ☐

14.3 Comments and hazards observed:

15. HAZARDS INTRODUCED BY OUTSIDE CONTRACTORS AND BUILDING WORKS

15.1 Is there satisfactory control over works carried out in the building by outside contractors (including "hot work" permits)? Yes ☐ No ☐

15.2 Are fire safety conditions imposed on outside contractors? Yes ☐ No ☐

15.3 If there are in-house maintenance personnel, are suitable precautions taken during "hot work", including use of hot work permits? N/A ☐ Yes ☐ No ☐

15.4 Comments:

FIRE PROTECTION MEASURES

16. MEANS OF ESCAPE FROM FIRE

16.1 It is considered that the building is provided with reasonable means of escape in case of fire. Yes ☐ No ☐

16.2 More specifically:

    Adequate provision of exits? Yes ☐ No ☐

    Exits easily and immediately openable where necessary? Yes ☐ No ☐

    Fire exits open in direction of escape where necessary? Yes ☐ No ☐

    Avoidance of sliding or revolving doors as fire exits where necessary? Yes ☐ No ☐

    Satisfactory means for securing exits? Yes ☐ No ☐

    Reasonable distances of travel:
- Where there is a single direction of travel? Yes ☐ No ☐
- Where there are alternative means of escape? Yes ☐ No ☐

    Suitable protection of escape routes? Yes ☐ No ☐

    Suitable fire precautions for all inner rooms? Yes ☐ No ☐

    Escape routes unobstructed? Yes ☐ No ☐

16.3 It is considered that the building is provided with reasonable arrangements for means of escape for disabled occupants. Yes ☐ No ☐

16.4 Comments and deficiencies observed:

17. MEASURES TO LIMIT FIRE SPREAD AND DEVELOPMENT

17.1 It is considered that there is:

    compartmentation of a reasonable standard[3]. Yes ☐ No ☐

    reasonable limitation of linings that may promote fire spread. Yes ☐ No ☐

17.2 Comments and deficiencies observed:

---

[3] Based on visual inspection of readily accessible areas, with a degree of sampling where appropriate.

18. ESCAPE LIGHTING

18.1 Reasonable standard of escape lighting system provided[4]?   Yes ☐   No ☐

18.2 Comments and deficiencies observed:

19. FIRE SAFETY SIGNS AND NOTICES

19.1 Reasonable standard of fire safety signs and notices?   Yes ☐   No ☐

19.2 Comments and deficiencies observed:

20. MEANS OF GIVING WARNING IN CASE OF FIRE

20.1 Reasonable manually operated electrical fire alarm system provided[5]?   Yes ☐   No ☐

20.2 Automatic fire detection provided?   Yes ☐ (throughout building)   Yes ☐ (part of building only)   No ☐

20.3 Remote transmission of alarm signals?   Yes ☐   No ☐

20.4 Comments and deficiencies observed?

21. MANUAL FIRE EXTINGUISHING APPLIANCES

21.1 Reasonable provision of portable fire extinguishers?   Yes ☐   No ☐

21.2 Hose reels provided?   Yes ☐   No ☐

21.3 Comments and deficiencies observed:

22. RELEVANT[‡] AUTOMATIC FIRE EXTINGUISHING SYSTEMS

22.1 Type of system:

22.2 Comments:

---

[4] Based on visual inspection, but no test of illuminance levels or verification of full compliance with relevant British Standard carried out.
[5] Based on visual inspection, but no audibility tests or verification of full compliance with relevant British Standard carried out.
[‡] Relevant to life safety and this risk assessment (as opposed purely to property protection).

PAS 79:2005

23. OTHER RELEVANT[‡] FIXED SYSTEMS

23.1 Type of system:

23.2 Comments:

MANAGEMENT OF FIRE SAFETY

24. PROCEDURES AND ARRANGEMENTS

24.1 Person responsible for fire safety[6]:

24.2 Competent person(s) available to assist in implementation of
fire safety legislation? Yes ☐ No ☐

Comments:

24.3 Appropriate fire procedures in place? Yes ☐ No ☐
(including arrangements for summoning the fire and rescue service)

Comments:

24.4 People nominated to respond to fire? N/A ☐ Yes ☐ No ☐

Comments:

24.5 People nominated to assist with evacuation? N/A ☐ Yes ☐ No ☐

Comments:

24.6 Appropriate liaison with fire brigade? N/A ☐ Yes ☐ No ☐

Comments:

[‡] Relevant to life safety and this risk assessment (as opposed purely to property protection).
[6] This is not intended to represent a legal interpretation of responsibility, but merely reflects the managerial arrangement in place at the time of this risk assessment.

PAS 79:2005

24.7 Routine in-house inspections of fire precautions (e.g. in the course of health and safety inspections)? N/A ☐ Yes ☐ No ☐

Comments:

25. TRAINING AND DRILLS

25.1 Are all staff given instruction on induction? Yes ☐ No ☐

Comments:

25.2 Are all staff given periodic "refresher training" at suitable intervals? Yes ☐ No ☐

Comments:

25.3 Are staff with special responsibilities (e.g. fire wardens) given additional training? N/A ☐ Yes ☐ No ☐

Comments:

25.4 Are fire drills carried out at appropriate intervals? Yes ☐ No ☐

Comments:

26. TESTING AND MAINTENANCE

26.1 Adequate maintenance of workplace? Yes ☐ No ☐

Comments and deficiencies observed:

26.2 Weekly testing and periodic servicing of fire detection and alarm system? Yes ☐ No ☐

Comments and deficiencies observed:

26.3 Monthly, six-monthly and annual testing routines for emergency lighting? Yes ☐ No ☐

Comments and deficiencies observed:

PAS 79:2005

26.4 Annual maintenance of fire extinguishing appliances? Yes ☐ No ☐

Comments and deficiencies observed:

26.5 Six-monthly inspection and annual testing of rising mains? N/A ☐ Yes ☐ No ☐

Comments and deficiencies observed:

26.6 Weekly testing and periodic inspection of sprinkler installations? N/A ☐ Yes ☐ No ☐

Comments:

26.7 Routine checks of final exit doors and/or security fastenings? N/A ☐ Yes ☐ No ☐

Comments:

26.8 Annual inspection and test of lightning protection system? N/A ☐ Yes ☐ No ☐

Comments:

26.9 Other relevant inspections or tests:

Comments:

27. RECORDS

27.1 Appropriate records of:

Fire drills? N/A ☐ Yes ☐ No ☐

Fire training? Yes ☐ No ☐

Fire alarm tests? N/A ☐ Yes ☐ No ☐

Escape lighting tests? N/A ☐ Yes ☐ No ☐

Maintenance and testing of other fire protection systems? N/A ☐ Yes ☐ No ☐

27.2 Comments:

PAS 79:2005

FIRE RISK ASSESSMENT

The following simple risk level estimator is based on a more general health and safety risk level estimator contained in BS 8800:

Potential consequences of fire ▶ Fire hazard ▼	Slight harm	Moderate harm	Extreme harm
Low	Trivial risk	Tolerable risk	Moderate risk
Medium	Tolerable risk	Moderate risk	Substantial risk
High	Moderate risk	Substantial risk	Intolerable risk

Taking into account the fire prevention measures observed at the time of this risk assessment, it is considered that the hazard from fire (probability of ignition) at this building is:

Low ☐   **Medium** ☐   **High** ☐

Taking into account the nature of the building and the occupants, as well as the fire protection and procedural arrangements observed at the time of this risk assessment, it is considered that the consequences for life safety in the event of fire would be:

Slight harm ☐   **Moderate harm** ☐   **Extreme harm** ☐

In this context, a definition of the above terms is as follows:

Slight harm: Outbreak of fire unlikely to result in serious injury or death of any occupant (other than an occupant sleeping in a bedroom in which a fire occurs).

Moderate harm: Outbreak of fire could result in injury of one or more occupants, but it is unlikely to involve multiple fatalities.

Extreme harm: Significant potential for serious injury or death of one or more occupants.

Accordingly, it is considered that the risk to life from fire at this building is:

Trivial ☐   Tolerable ☐   Moderate ☐   Substantial ☐   Intolerable ☐

A suitable risk-based control plan should involve effort and urgency that is proportional to risk. The following risk-based control plan is based on one advocated by BS 8800 for general health and safety risks:

Risk Level	Action and timescale
Trivial	No action is required and no detailed records need be kept.
Tolerable	No major additional controls required. However, there may be a need for consideration of improvements that involve minor or limited cost.
Moderate	It is essential that efforts are made to reduce the risk. Risk reduction measures should be implemented within a defined time period. Where moderate risk is associated with consequences that constitute extreme harm, further assessment may be required to establish more precisely the likelihood of harm as a basis for determining the priority for improved control measures.
Substantial	Considerable resources may have to be allocated to reduce the risk. If the building is unoccupied, it should not be occupied until the risk has been reduced. If the building is occupied, urgent action should be taken.
Intolerable	Building (or relevant area) should not be occupied until the risk is reduced.

Please note that, although the purpose of this section is to place the fire risk in context, the above approach to fire risk assessment is subjective and for guidance only. All hazards and deficiencies identified in this report should be addressed by implementing all recommendations contained in the following section. The risk assessment should be reviewed periodically.

ACTION PLAN

It is considered that the following recommendations should be implemented in order to reduce fire risk to, or maintain it at, the following level:

Trivial ☐   Tolerable ☐

Definition of priorities (where applicable):

1   Priority
                                                                                                           (where
                                                                                                           applicable)

Annex B (informative)
Fire hazard prompt list

B.1 This Annex sets out, in Table B.1, a list of fire hazards that are normally considered in the fire risk assessment. Typical key measures for the elimination or control of each hazard are given, along with some relevant codes of practice or guidance documents.

B.2 This prompt list is not necessarily exhaustive, particularly in respect of measures for control and elimination of fire hazards, and there might be a need to consider further hazards and measures to prevent fire in the course of the fire risk assessment, particularly if work processes give rise to more specific fire hazards. Similarly, the codes of practice and guidance documents referenced are intended only to comprise a representative sample of those available.

Table B.1 — Fire hazards, elimination or control measures and relevant codes of practice

Fire hazard	Typical key measures for control or elimination of the fire hazard	Relevant code of practice or guidance document
Electrical faults	Periodic inspection and testing of fixed electrical installation. Portable appliance testing. Suitable control over employees' and visitors' use of their own electrical appliances. Limitation of trailing leads and adaptors.	IEE Guidance Note 3 [9]. IEE *Code of practice for in-service inspection and testing of electrical equipment* [10]. HSE HSG 107 [11]
Smoking	Prohibition or limitation of smoking in either the entire building or in appropriate areas of the building. Suitable arrangements for those who wish to smoke.	
Arson	Basic security measures to prevent malicious ignition by outsiders. Avoidance of unnecessary fire load in close proximity to the building or available for ignition by outsiders.	*The prevention and control of arson* [12].
Improper use of portable heaters	Avoidance of use of portable heaters as far as practicable. If portable heaters are used, avoidance of the most hazardous types of heater. Suitable measures to minimize the likelihood of ignition of combustible materials.	
Faults in fixed heating installations	Regular maintenance of installations.	
Use of cooking appliances	Suitable design of kitchens. Availability of suitable fire extinguishing appliances to deal with small fires. Regular replacement of grease filters and cleaning of extract ductwork.	*Recommendations for cooking equipment (other than fish and chip shop frying ranges)* [13]. *Fire risk assessment in catering ventilation* [14].
Lightning	Provide lightning protection system if likelihood of lightning strike warrants it.	BS 6651.
Contractors' operations and "hot work" by maintenance staff	Suitable fire safety conditions in contracts with outside contractors. Suitable control over outside contractors while in the building. Suitable control over hazardous activities by in-house maintenance personnel, such as "hot work" involving cutting, welding, use of blowlamps, etc.	*Standard fire precautions for contractors engaged on Crown works* [15]. *Fire prevention on construction sites* [16]. *Fire safety in construction work* [17].
Poor housekeeping and inadequate control over general fire hazards or specific fire hazards associated with work activities	Separation of combustible materials from ignition sources. Avoidance of unnecessary accumulation of combustible materials or waste. Appropriate storage of hazardous materials. Avoidance of inappropriate storage of combustible materials. Proper maintenance of the workplace. Routine safety inspections.	There are numerous publications on the subject of fire prevention. Most publications on the subject give guidance on housekeeping and maintenance issues. More specific guidance exists for numerous occupancies, work processes and related hazards. See, for example *Fire precautions – A guide for management* [18], and a wide range of publications produced by the Fire Protection Association (www.thefpa.co.uk/pdf/pub_cat.pdf).

Annex C (normative)
Key factors to consider in assessment of means of escape

C.1 Table C.1 shows the key factors that should always be explicitly considered in assessment of means of escape. Most of the factors are quite broad and encompass a number of more specific issues. These key factors can be used as a form of prompt list and should, therefore, normally be shown in the documented fire risk assessment (see Clause **9**).

C.2 The more specific issues should always be considered in the fire risk assessment process but may or may not be explicitly shown in the documented fire risk assessment. Where the experience of the fire risk assessor is limited, it might be of value for at least some of the specific issues to be included in the pro-forma used, so that they act as prompts or reminders to the fire risk assessor.

C.3 Where it is determined that there are significant departures in compliance of any key factor or specific issue with recognized codes of practice, but it is considered that the departures are acceptable (and, hence, no relevant recommendation needs to be made in the action plan), the reasoning behind the acceptance of each departure should be documented in the fire risk assessment (see **9.2 b)**).

Table C.1 — Key factors and specific issues to consider in means of escape

Key factor	Specific issues to consider	Notes
Design of escape routes	<ul><li>Do escape routes lead to final exits?</li><li>Do doors on means of escape open in the direction of escape where necessary?</li><li>Will occupants of inner rooms (see **3.56**) be aware of a fire in the access rooms?</li><li>Do revolving doors or sliding doors have suitable by-pass doors where necessary?</li><li>Are there (and is there a need for) alternative escape routes (see **3.3**)?</li></ul>	
Distances of travel	<ul><li>Are travel distances (see **3.80**) reasonable?</li><li>Are travel distances in dead ends (see **3.14**) suitably limited?</li></ul>	Recommended maximum travel distances are given in all codes of practice on means of escape, but these figures should not be considered in isolation of other fire protection measures (see **14.1.3**). The likely rate of fire development, and the consequent time available for escape, need to be taken into account.
Protection of escape routes	<ul><li>Are escape routes, such as staircases, dead end corridors, bedroom corridors, etc, protected (see **3.71**) where necessary?</li><li>Are all fire resisting doors properly self-closing, kept locked shut or only held open by suitable automatic door release mechanisms (see **3.4**)?</li></ul>	Where automatic door release mechanisms are used, it is important to ensure that there is adequate provision of suitably sited smoke detectors.
Adequate provision of exits and escape routes	<ul><li>Is there a sufficient number of fire exits and escape routes?</li><li>Are the number and widths of fire exits and escape routes sufficient for the number of occupants?</li></ul>	Methods of calculating exit capacity are given in all codes of practice on means of escape
Exits easily and immediately openable	<ul><li>Are fire exits easily openable without, for example, the use of a key?</li><li>Is there only a single means of securing each fire exit?</li><li>Where necessary, do the means of securing fire exits comprise panic bolts (see **3.66**) or panic latches (see **3.67**)?</li><li>Where electronic locking is used, is its use acceptable, and are the means of releasing the locks suitable?</li></ul>	
Escape routes unobstructed	<ul><li>Are escape routes kept unobstructed?</li><li>Are adequate widths of corridors and other escape routes maintained at all times?</li></ul>	Escape route widths should be sufficient for the number of people who need to use the escape route.

Annex D (informative)
Model pro-forma for documentation of a review of an existing fire risk assessiment

D.1 This Annex contains a pro-forma for documentation of a review of an existing fire risk assessment. If the pro-forma is completed by a competent person, the format and scope of the review will be suitable and sufficient to satisfy the recommendations in Clause **19**.

D.2 The format of the documented review may vary from that shown in this Annex, provided the recommendations in Clause **19** are satisfied. For example, the level to which principal issues are broken down into their component factors may vary, provided it is clear that the principal issues addressed in the original fire risk assessment have been addressed, or that the scope of the review is limited to, for example, a material alteration that has resulted in the review (see Clause **19**).

PAS 79:2005

WORKPLACE FIRE PRECAUTIONS LEGISLATION

PERIODIC REVIEW OF FIRE RISK ASSESSMENT

Address of Property:

Person(s) Consulted:

Assessor:

Date of Fire Risk Assessment:

Date of Previous Fire Risk Assessment:

Suggested Date for Review[7]:

The purpose of this report is to provide an assessment of the risk to life from fire in these buildings, and, where appropriate, to make recommendations to ensure compliance with fire safety legislation. The report does not address the risk to property or business continuity from fire.

[Date]

---

[7] The original fire risk assessment should be reviewed again by a competent person by the date indicated above or at such earlier time as there is reason to suspect that it is no longer valid or there have been significant changes.

PAS 79:2005

GENERAL INFORMATION

1. Significant changes identified since the time of the previous fire risk assessment in respect of:

1.1 The premises:

1.2 The occupancy:

1.3 The occupants (including occupants at special risk):

1.4 Fire loss experience:

1.5 Application of fire safety legislation:

1.6 Other relevant information:

FIRE HAZARDS AND THEIR ELIMINATION OR CONTROL

2. Significant changes in measures to prevent fire since the time of the fire risk assessment:

3.1 Are there adequate measures to prevent fire? Yes ☐ No ☐

3.2 Comments and hazards observed:

4.1 Are housekeeping and maintenance adequate? Yes ☐ No ☐

4.2 Comments and deficiencies observed:

PAS 79:2005

FIRE PROTECTION MEASURES

5.1 Significant changes in fire protection measures since the time of the fire risk assessment:

6.1 Are the means of escape from fire adequate?   Yes ☐   No ☐

6.2 Comments and deficiencies observed:

7.1 Are compartmentation and linings satisfactory?   Yes ☐   No ☐

7.2 Comments and deficiencies observed:

8.1 Is there reasonable emergency escape lighting[8]?   Yes ☐   No ☐

8.2 Comments and deficiencies observed:

9.1 Are there adequate fire safety signs and notices?   Yes ☐   No ☐

9.2 Comments and deficiencies observed:

10.1 Are the means of giving warning of fire adequate[9]?   Yes ☐   No ☐

10.2 Comments and deficiencies observed:

11.1 Is the provision of fire extinguishing appliances adequate?   Yes ☐   No ☐

11.2 Comments and deficiencies observed:

12.1 Comments on other fixed fire protection systems?

---

[8] Based on visual inspection only.

PAS 79:2005

MANAGEMENT OF FIRE SAFETY

13.1 Significant changes in management of fire safety since the time of the fire risk assessment:

14.1 Are arrangements for management of fire safety adequate?   Yes ☐   No ☐

14.2 Comments and deficiencies observed:

15.1 Are fire procedures adequate?   Yes ☐   No ☐

15.2 Comments and deficiencies observed:

16.1 Are the arrangements for staff training and fire drills adequate?   Yes ☐   No ☐

16.2 Comments and deficiencies observed:

17.1 Are the arrangements for testing and maintenance of fire protection systems and equipment adequate?   Yes ☐   No ☐

17.2 Comments and deficiencies observed:

18.1 Are there adequate records of testing, maintenance, training and drills?   Yes ☐   No ☐

18.2 Comments and deficiencies observed:

FIRE RISK ASSESSMENT

On the basis of the criteria set out in the original fire risk assessment, it is considered that the current risk to life from fire at these premises is:

Trivial ☐   Tolerable ☐   Moderate ☐   Substantial ☐   Intolerable ☐

PAS 79:2005

ACTION ON PREVIOUS ACTION PLAN

Have all previous recommendations been satisfactorily addressed?

Yes ☐   No ☐

Brief details of recommendations not yet implemented.

1

NEW ACTION PLAN

It is considered that the following recommendations should be implemented in order to reduce fire risk to, or maintain it at, the following level:

Trivial ☐   Tolerable ☐

Definition of priorities (where applicable):

1   Priority (where applicable)

Bibliography

Standards publications

BS 4422-1:1987, *Glossary of terms associated with fire – Part 1: General terms and phenomena of fire.*

BS 5266-1:1999, *Emergency lighting – Part 1: Code of practice for the emergency lighting of premises other than cinemas and certain other specified premises used for entertainment.*

BS 5266-7:1999 (BS EN 1838:1999), *Lighting applications – Part 7: Emergency lighting.*

BS 5306-1:1976 (1988), *Fire extinguishing installations and equipment on premises – Part 1: Hydrant systems, hose reels and foam inlets.*

BS 5306-2:1990, *Fire extinguishing installations and equipment on premises – Part 2: Specification for sprinkler systems.*

BS 5306-8:2000, *Fire extinguishing installations and equipment on premises – Part 8: Selection and installation of portable fire extinguishers – Code of practice.*

BS 5499-4:2000, *Safety signs, including fire safety signs – Part 4: Code of practice for escape route signing.*

BS 5588 (all parts), *Fire precautions in the design, construction and use of buildings.*

BS 5839-1:2002, *Fire detection and fire alarm systems for buildings – Part 1: Code of practice for system design, installation, commissioning and maintenance.*

BS 5839-6: 2004, *Fire detection and fire alarm systems for buildings – Part 6: Code of practice for the design, installation and maintenance of fire detection and fire alarm systems in dwellings.*

BS 6651:1999, *Code of practice for protection of structures against lightning.*

BS 7671:2001, *Requirements for electrical installations – IEE Wiring Regulations – Sixteenth edition.*

BS 7974:2001, *Application of fire safety engineering principles to the design of buildings.*

BS 8800:1996, *Guide to occupational health and safety management systems.*

Other documents

[1] GREAT BRITAIN. Management of Health and Safety at Work Regulations 1999. London: The Stationery Office (TSO).

[2] GREAT BRITAIN. Fire Precautions (Workplace) Regulations 1997 (as amended). London: The Stationery Office (TSO).

[3] NORTHERN IRELAND. Management of Health and Safety at Work Regulations (Northern Ireland) 2000. Belfast: The Stationery Office (TSO).

[4] NORTHERN IRELAND. Fire Precautions (Workplace) Regulations (Northern Ireland) 2001. Belfast: The Stationery Office (TSO).

[5] GREAT BRITAIN. Regulatory Reform Act 2001. London: The Stationery Office (TSO).

[6] *Fire Safety. An Employer's Guide*. Home Office, Scottish Executive, Health and Safety Executive and Department of Environment (Northern Ireland). London: The Stationery Office (TSO).

[7] The Building Regulations (England and Wales). 1991. *Approved Document B. Fire Safety*. 2000. London: The Stationery Office.

[8] GREAT BRITAIN. Health and Safety (Safety Signs and Signals) Regulations 1996. London: The Stationery Office.

[9] *Guidance Note 3: Inspection and Testing, 4th Edition*. 2002. London: Institution of Electrical Engineers.

[10] *Code of practice for in-service inspection and testing of electrical equipment*. 2001. London: Institution of Electrical Engineers.

[11] HEALTH AND SAFETY EXECUTIVE. HSG 107. *Maintaining portable and transportable electrical equipment*. 2000. London: The Stationery Office.

[12] *The prevention and control of arson*. 1999. London: Fire Protection Association.

[13] *Recommendations for cooking equipment (other than fish and chip shop frying ranges)*. Insurers' Fire Research Strategy Funding Scheme (InFiReS). 2003. London. Fire Protection Association.

[14] *Fire risk assessment in catering ventilation*. Association of British Insurers and BSRIA. 2001. Bracknell. BSRIA.

[15] *Standard fire precautions for contractors engaged on Crown works applicable to contractors engaged on works for Crown Civil and Defence Estates*. 1995. London: The Stationery Office.

[16] *Fire prevention on construction sites*. 2000. London: Fire Protection Association.

[17] HEALTH AND SAFETY EXECUTIVE. HSG 168. *Fire safety in construction work*. 1997.

[18] TODD, C.S. *Fire precautions. A guide for management*. 2000. London: Gower Publishing Limited.

[19] *Ensuring Best Practice*. 2004. London: Association for Specialist Fire Protection. www.asfp.org.uk

[20] Building Standards (Scotland) Regulations 1990. Technical standards. 2001. London: The Stationery Office